POCKET GUIDE TO
INSECTS

First published in 2014 by Bloomsbury Publishing Plc,
50 Bedford Square, London WC1B 3DP

www.bloomsbury.com

UK ISBN (print) 978-1-47290-915-2

Bloomsbury Publishing, London, New Delhi, New York and Sydney

Bloomsbury is a trademark of Bloomsbury Publishing Plc

A CIP catalogue record for this book is available from the British Library
Library of Congress Cataloging-in-Publication Data has been applied for

Publisher: Nigel Redman
Project editor: Alice Ward
Design by Rod Teasdale

Printed in China by Toppan Leefung Printing Co Ltd.

This book is produced using paper that is made from wood grown in managed sustainable forests. It is natural, renewable and recyclable. The logging and manufacturing processes conform to the environmental regulation of the country of origin.

10 9 8 7 6 5 4 3 2 1

POCKET GUIDE TO
INSECTS

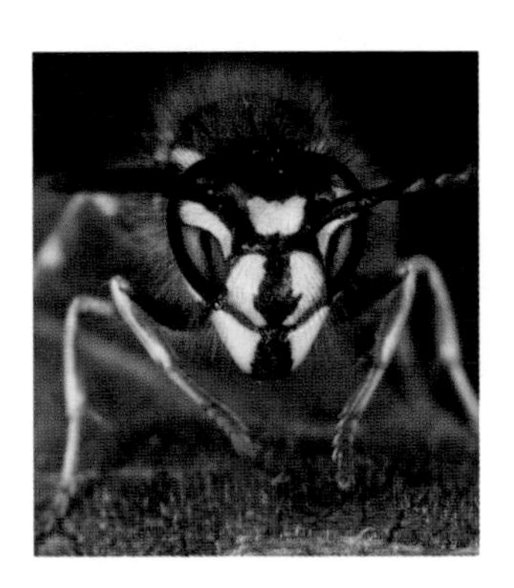

Bob Gibbons

B L O O M S B U R Y
LONDON • NEW DELHI • NEW YORK • SYDNEY

CONTENTS

INTRODUCTION

There are about 24,000 known species of insects in the British Isles, in a vast variety of forms, so you might reasonably wonder what use a little book on insects like this might be. In fact, the great majority of insects are quite small, often barely noticed, and hard to identify without specialised knowledge and equipment. For this book about 240 easily noticed species (including secondary species, with only a short entry) have been selected from a range of orders and families. They can all be identified reasonably easily in the field. No attempt has been made to cover the range of species for butterflies and moths, as both groups are fully covered in other books, including a forthcoming one in this series.

The geographical area covered is essentially the British Isles, including Ireland, but with a nod to the nearby Continent. A few non-British species are included, either because they are commonly seen in the adjacent parts of continental Europe, or because they are likely to occur here in the near future.

HOW TO USE THIS BOOK

This is essentially a photographic guide, without keys, so the best starting point is to look through the photographs to find something similar to what you have seen. The information for the larger orders, such as flies and beetles, includes an introduction to help confirm the general identity of your species. If it does not quite fit, check nearby pages and also any similar species mentioned at the end of the species account. It is also worth looking at the information given on size, normal habitats, distribution in the region and flight period to try to confirm an identification. The following information is provided under separate headings within the acounts.

Flight period gives an indication of when the adult insect is likely to be seen. Some insects do not actually fly – this is just a convenient term for the adult period. Bear in mind that in particularly warm or cool sites, or in unusual years, the adults may be seen outside the given period.

Habitat and distribution gives the normal habitats of the species, for example woodland, but of course many species are quite mobile and turn up in other habitats, especially if they are attracted by lights at night. Some details of where the species occur in Britain

Perfect insect habitat – flowery, sheltered and bushy.

and Ireland are given, but they are usually simplifications, and many species are spreading or declining, so this information should not be considered as a defining feature.

Similar species gives an indication either of species that might be confused with the main one, or of some closely related species. In many cases there are a great many rather similar species, and in these instances it is hard to say anything useful in a small space.

WHAT IS AN INSECT?

Insects are part of the enormous invertebrate group known as the arthropods, which includes many non-insect animals such as spiders, harvestmen, millipedes and others. The main characteristic of the group is the possession of a hard external shell or skeleton, with flexible joints that allow the animal to move. Insects are defined within the group by some or all of these features:

1. The presence of wings. No arthropods other than insects have wings, although not all insects have wings.

2. Insects normally have three pairs of legs. Although some insects have fewer legs, the three pairs are usually visible at some point in the life-cycle, or there is at least an indication of where they should be. They never have more than three pairs of legs.

THE BODY DESIGN OF A COMMON EARWIG

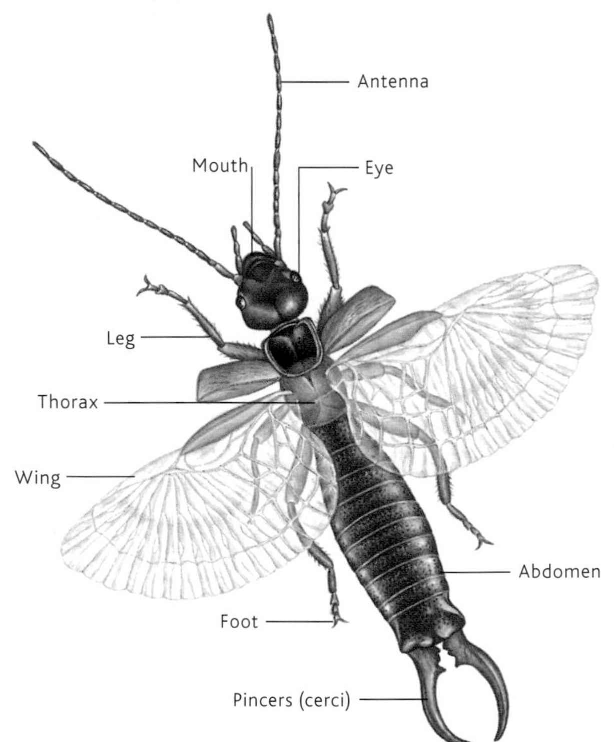

Fig. 1

3. Insect bodies are usually clearly divided into a head, thorax and abdomen. There are usually two antennae on the head, and the thorax bears the legs and wings if present. Most winged insects have two pairs of wings, but the diptera (see p. 90) have only one pair, with the hind pair being replaced by a couple of tiny balancing organs called halteres. The general structure of a typical insect – a Common Earwig – is shown in fig. 1. Other insects differ in shape and details, but most of these parts are found on all insects. See the glossary (p. 186) for descriptions of some other insect parts.

FINDING INSECTS

Although insects are everywhere, it can be surprisingly difficult to find specific species, or groups of insects such as dragonflies, without local knowledge. In general the best way to discover a range of interesting and attractive insects is by visiting high-quality semi-natural habitats, such as chalk grassland (especially in June and July), old woodland (especially in spring), heathland (especially in summer) and wetlands such as bogs, pools and lakes. All of these will be rich in insect life unless badly polluted.

Nature reserves frequently preserve the best examples of these habitats, and most are open to the public. Some examples of organisations that have reserves are given in the Organisations and Societies section (see p. 188). It is also worth searching the internet as there are many websites that give details of good sites for particular species or groups.

When out in the field it is worth walking slowly and scanning ahead, trying to see the larger insects before they see you. If you approach slowly they will often not see you as a threat and may stay put.

INSECT CONSERVATION

Insects are declining alarmingly in most developed countries. The combination of habitat loss, habitat fragmentation and widespread use of pesticides and other damaging chemicals has led to huge losses of numbers, and many species have become extinct in recent decades. Tragic in itself, this loss also affects our populations of birds, bats and other animals that are so dependent on insects.

It is little wonder that insect conservation has lagged behind that of other major groups – insects are small and often inconspicuous, there is an overwhelming number of species and insects are often perceived as damaging. We also frequently know little about what insect species need to flourish. Yet they are a really vital part of the web of natural life, and more needs to be done. Several organisations listed (see p. 188) are specifically concerned with insect conservation, and of course the general nature conservation organisations help by protecting good habitats. Join as many of these as you can, and get involved.

Footnote: *An asterisk * denotes that the species does not regularly occur in the UK.*

MAYFLIES, EPHEMEROPTERA

A group of about 50 species in Britain, the mayflies are an interesting primitive order of insects with some unusual features. Eggs are laid by the female on the water's surface, and these hatch into nymphs that undergo many moults before emerging as winged insects. This is the only insect group in which the fully winged insect is not the adult. The winged stages are known as duns, and do not have full adult colours or mature sexual organs; shortly after, they moult again into the fully coloured adult insects, which are known as spinners. In most species these adults live for a very short time – sometimes just for a few hours – and they do not feed. They exist only to mate and disperse.

The bodies of mayflies are soft and slightly flattened, and there are two or three long 'tails' at the rear of the abdomen and very short antennae. The front legs of males are particularly long. There are two pairs of delicate wings, but the hindwings are much smaller than the front wings and may even be absent altogether.

Drake Mackerel Mayfly

Ephemera vulgata

This is one of the larger and more distinctive mayflies, with boldly dark-patterned wings, a long body up to 22mm long, two long 'tails' and a patterned orange-brown abdomen with triangle markings. Its unusual common name comes from the fishing fraternity, which has given specific names to many insects that fish feed on.

FLIGHT PERIOD April–August.

HABITAT AND DISTRIBUTION Widespread in England and Wales, though more common in the east, in still or slow-moving water bodies with a silty base.

SIMILAR SPECIES *E. danica* is very similar in size and shape, but has three tails, is rather paler and lacks the triangle markings.

Right: Male Drake Mackerel – note the long front legs.

DAMSELFLIES AND DRAGONFLIES, ODONATA

These are two closely related groups of predatory insect, all with an aquatic stage in their life-cycle. Damselflies (suborder Zygoptera) are more slender, delicate insects than dragonflies. They normally hold their wings closed above the abdomen (or sometimes half open) when at rest. Their eyes are well separated and located on either side of the head. Dragonflies (suborder Anisoptera) are more robust, very strong-flying insects that hold their wings out widely on each side when at rest. Their eyes are large and normally touch in the centre of the head (except in the clubtails).

All species are most likely to be found in wet habitats, though dragonflies in particular hunt widely away from water. About 26 species of damselfly and 40 species of dragonfly occur in the area.

A male Black Darter dragonfly perched.

Damselflies

Banded Demoiselle or Banded Agrion
Calopteryx splendens

This is one of the most familiar and distinctive of damselflies, though it is sometimes mistaken for a butterfly due to its broad and colourful wings. It is about 40mm long. Males are metallic blue, with a bold blue band on each wing occupying about half of the wing area. Females are metallic green, with clear, pale greenish wings. They fly with a floppy, butterfly-like flight, perching frequently with closed wings on waterside vegetation.

FLIGHT PERIOD May–August.

HABITAT AND DISTRIBUTION Most frequent along larger slow-flowing rivers, avoiding fast-flowing, cold or acid rivers. Widespread and common in southerly areas, as far north as northern England, and Northern Ireland.

SIMILAR SPECIES Beautiful Demoiselle *C. virgo* is similar in size and shape, but males have wings that are virtually all smoky brownish-blue, while females have pale brownish-green wings. Similar distribution and habitats, though more likely to be found in faster-flowing, more acid or more shady streams.

A male Banded Agrion damselfly.

Common Spreadwing or Common Emerald Damselfly

Lestes sponsa

The spreadwings are notable among the damselflies for usually settling with their wings spread roughly half open (hence their name), and are readily identifiable as a group. This species is the most common of the spreadwings. It is a medium-sized damselfly with a metallic green thorax and abdomen, all green in females, but with a blue 'tail' and abdominal patch in males.

FLIGHT PERIOD May–October.

HABITAT AND DISTRIBUTION Widespread throughout Britain and northern Europe in wet areas with still water.

SIMILAR SPECIES Several similar species occur, needing close examination for identification. Migrant Spreadwing *L. barbarus* is perhaps most distinctive, with a two-coloured pterostigma, pale colour and white-tipped abdomen. It occurs occasionally in Britain, regularly as far north as northern Germany and sporadically further north.

Male Common Spreadwing in typical pose with wings half open.

Common Bluetail

Ischnura elegans

This is a common small damselfly with an abdomen 20–30mm in length. Males are dark blackish-bronze with a bright blue single penultimate abdominal segment (segment eight), which appears like a little blue tail-light. Females are similar, though more variable in colour. Apart from in the species noted below, this is a distinctive combination.

FLIGHT PERIOD May–September.

HABITAT AND DISTRIBUTION Widespread and common throughout Britain and northern Europe in wet areas with still water; less common in running water.

SIMILAR SPECIES Small Bluetail *I. pumilio* is similar but smaller, and the blue 'tail-light' covers the whole of segment nine and just a little of segment eight. It is primarily a southern species and rather rare in the area, with scattered fluctuating populations in southern Britain and Ireland, and northwards on mainland Europe to Denmark.

Male common Bluetail.

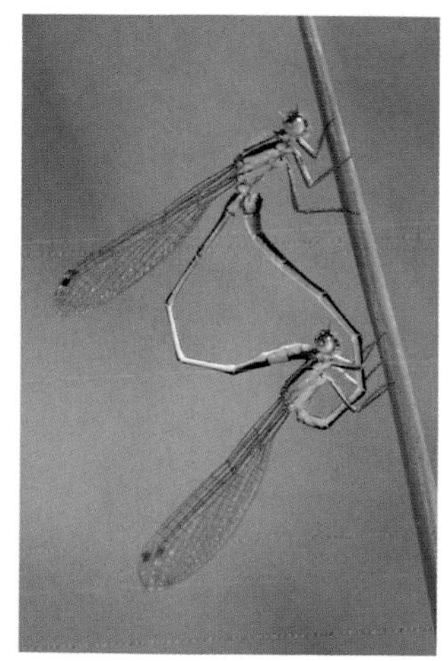

Common Bluetail mating pair.

Common Bluet
Enallagma cyathigera

This is almost certainly the most common damselfly in the area, resembling many other species in its blue and black pattern. The best identification features are the presence of only one black stripe on each side of the blue thorax and, in males, a little black, stalked 'ball' marking on segment two of the abdomen. Females are less distinctive and more variable.

FLIGHT PERIOD May–September.

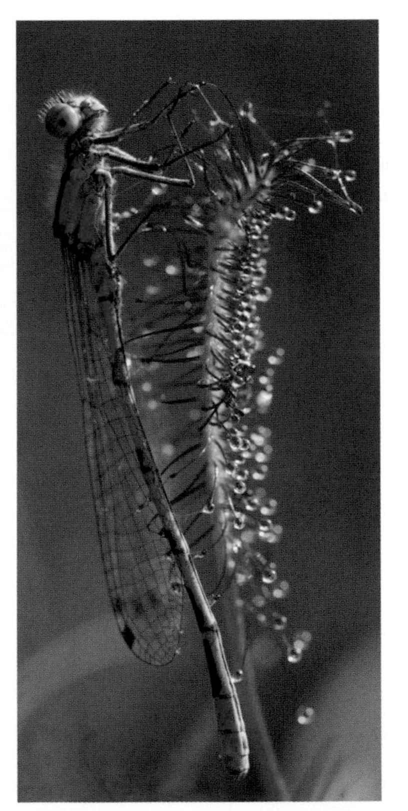

Male Common Bluet trapped on sundew.

HABITAT AND DISTRIBUTION Occurs throughout the whole area in a variety of waters, especially acidic lakes and ponds, and is frequently abundant.

SIMILAR SPECIES Many *Coenagrion* species (see opposite) are similar, but all males have an additional short black stripe on the blue thorax, as well as individual differences.

Azure Bluet

Coenagrion puella

A typical 'blue damselfly', Azure Bluet is more slender and longer than Common Bluet, with the typical *Coenagrion* extra short black stripe on the side of the male's thorax. Males have a rather distinctive 'open-box' black marking on segment two of the abdomen (compare with Common Bluet, opposite). Females are usually green and black.

FLIGHT PERIOD May–September.

HABITAT AND DISTRIBUTION Common throughout most of Britain except the north, in a variety of waters, especially if they are well vegetated.

SIMILAR SPECIES Mercury Bluet *C. mercuriale* is a rare southern species with a distinctive winged mercury mark on segment two; Spearhead Bluet *C. hastulatum* is the most common of the group in far northern Europe, distinguished by the 'stalked spearhead' marking on segment two. Within Britain it is found only in Scotland.

Mating Azure Bluets, male above, female laying.

Large Redeye or Red-eyed Damselfly

Erythromma najas

This very distinctive medium-sized damselfly is (with the Small Redeye) the only red-eyed damselfly in northern Europe. Males are bluish-black with a blue tail (segments 9–10) and striking red eyes; females are very dark with no blue tail, and duller reddish eyes.

FLIGHT PERIOD May–August.

HABITAT AND DISTRIBUTION Locally common in well-vegetated still or slow-moving waters with floating aquatic plants, and virtually confined to England within the UK.

SIMILAR SPECIES Small Redeye *E. viridulum* is smaller, and in males the blue tail extends along the sides of segment eight and on segments adjacent to the thorax. It is a more southern species, but is steadily spreading and is found in similar habitats.

Male Large Redeye on leaf.

Large Red Damsel
Pyrrhosoma nymphula

This is a common and distinctive damselfly in much of northern Europe, where it is frequently the first damselfly species to appear in spring. It is of medium size with a largely red abdomen that becomes black towards the tip (though some females may be almost all black from above), and black legs. The thorax is black above with red stripes.

FLIGHT PERIOD April–August.

HABITAT AND DISTRIBUTION Widespread and common, most frequently around still, well-vegetated waters, throughout Britain.

SIMILAR SPECIES Small Red Damsel *Ceriagrion tenellum* is smaller with red legs, and males have a wholly red abdomen. It is rarer, occurring from Wales and northern Germany southwards in bogs and seepages.

Male Large Red Damsel perched.

Blue Featherleg or White-legged Damselfly

Platycnemis pennipes

Males of this species are distinctive and attractive, with pale blue-white, non-metallic bodies tipped with black, and broad, white feathery legs edged with long bristles. They also have unusually broad heads with widely separated eyes. Females are paler, lacking the black on the last few abdominal segments.

FLIGHT PERIOD May–August.

HABITAT AND DISTRIBUTION A largely southern species reaching as far north as central England. It is locally common, usually in slow-flowing rivers and streams, but also in still waters, especially in the southern part of its range.

SIMILAR SPECIES None.

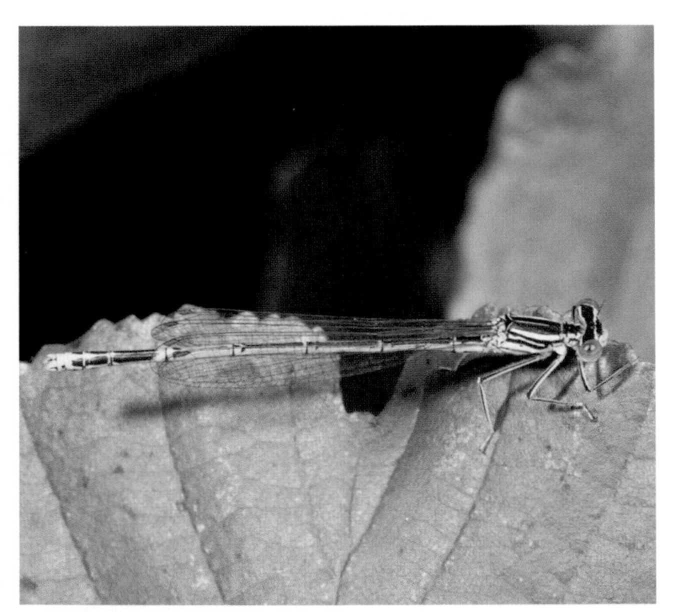

Male Blue Featherleg.

Dragonflies

Blue Emperor

Anax imperator

Blue Emperor is a very large dragonfly, both in length (up to 85mm long) and in general bulk. Males are particularly distinctive, with a bright blue abdomen marked only with a black line, and a bright green thorax. Females are less conspicuous – largely greenish with a black stripe. The insects patrol strongly over water, flying with a slight downward-curved abdomen.

FLIGHT PERIOD June–September

HABITAT AND DISTRIBUTION An essentially southern species found throughout most of Europe, reaching northwards as far as northern England, with occasional records further north. It is currently expanding its range, and occurs most frequently on large, well-vegetated still water bodies.

SIMILAR SPECIES Lesser Emperor *A. parthenope* is smaller and generally browner, with a distinct blue 'saddle' on the first segments of the abdomen. It is more strongly southern and found in similar habitats, but is gradually spreading northwards. It is occasional in Britain.

A male Blue Emperor dragonfly perched.

Blue Hawker or Southern Hawker

Aeshna cyanea

The hawkers are among the largest dragonflies. They are up to almost 80mm long, with a strong, aggressive flight and active habits. Blue Hawker is the most common hawker in much of the area. Its key features (in the male) include a blackish ground colour, with two broad apple-green stripes on the thorax and paired green spots along the abdomen that become blue on the last three segments and fused on the last two. Females are less boldly marked. It flies strongly, with a slightly drooping abdomen, often well away from water and into late evening.

FLIGHT PERIOD Mid June–October.

HABITAT AND DISTRIBUTION Widespread in southern Britain, rare in Ireland. It breeds in small, shady ponds, but may be seen almost anywhere.

Female Blue Hawker perched.

Male Blue Hawker in flight.

SIMILAR SPECIES Migrant Hawker *A. mixta* is smaller, lacks the green colouration and has just two small yellow spots on the thorax. It is much less territorial, and many individuals may be seen together. It is widespread as both a resident and a migrant. Brown Hawker *A. grandis* is almost all brown, very large and has brown-tinted wings – a distinctive combination. It is common almost throughout the area except the far north, in still or slow-flowing large water bodies. Moorland Hawker *A. juncea* is similar to Blue Hawker, but with thin yellow thorax stripes, a yellow front edge to the wing and separated blue spots on all the abdomen segments. It is the most common species in northern and upland areas, and is absent from many lowland areas.

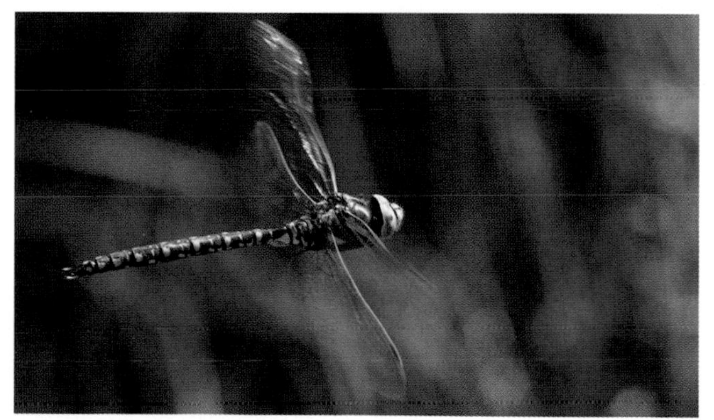

Male Moorland Hawker in flight.

Hairy Hawker

Brachytron pratense

Although this species resembles the *Aeshna* hawkers, it is not that closely related to them. It differs in being smaller than most of them (up to about 60mm long), with a noticeably downy body. There are two green thoracic stripes and paired blue spots all down the abdomen (like in Moorland Hawker). The wings each have a very long, narrow brownish pterostigma. The species' early flight season also usually serves to separate it from the *Aeshna* species.

FLIGHT PERIOD April–July, peaking May–June.

HABITAT AND DISTRIBUTION Scattered in southern Britain, including most of England. It is particularly common around the ditches of coastal grazing marshes, although by no means confined to them.

SIMILAR SPECIES Most likely to be confused with the *Aeshna* hawkers.

A male Hairy Hawker.

Common Clubtail

Gomphus vulgatissimus

The clubtails are a distinctive, mainly southern group within Europe, with separated eyes, a yellow line down the abdomen and a thickened 'tail' to the abdomen in most species. Common Clubtail is much the most widespread of the group in northern Europe. It is larger and darker than most similar species, with black legs, greenish eyes and a largely black, strongly clubbed abdomen.

FLIGHT PERIOD Late May–early August.

HABITAT AND DISTRIBUTION Scattered in southern Britain; most common in eastern Europe. It favours large, sandy-bottomed rivers in the lowlands, and is found less commonly in lakes and ponds.

SIMILAR SPECIES *Western Clubtail *G. pulchellus* is more slender, yellower and barely club-tailed, with blue eyes. It is endemic to south-west Europe, extending as far north as Germany.

Perched female Common Clubtail.

Common Goldenring or Golden-ringed Dragonfly

Cordulegaster boltonii

A large, conspicuous and distinctive dragonfly up to 85mm long, Common Goldenring has a boldly marked black body, with yellow rings along the abdomen and yellow stripes on the thorax. The eyes are large and greenish, and just touch at a single point. The sexes are similar, except that the female is longer due to a projecting, spike-like scale at the tip of her body, which is used to assist egg-laying on stream bottoms.

FLIGHT PERIOD Late May–late August.

HABITAT AND DISTRIBUTION Scattered in western and upland areas where there is suitable habitat, but absent from much of the lowlands and all of Ireland. Almost confined to well-oxygenated, clean small rivers and streams, which are often partially shaded.

SIMILAR SPECIES None in the UK.

Male Common Goldenring on riverside perch.

Brilliant Emerald

Somatochlora metallica

Brilliant Emerald is a very attractive, shiny metallic green, medium-sized dragonfly up to 55mm long. The thorax and abdomen are predominantly shiny green, with just a few yellow patches on the upper side at the base of the abdomen in males. The wings, especially in males, are often yellowish-brown at the base. Males patrol tirelessly around lakes, usually keeping close to the edges, often in shade.

FLIGHT PERIOD May–late August.

HABITAT AND DISTRIBUTION Common and widespread over most of northern Europe; rare in the UK, where it is confined to south-east England and the Scottish Highlands.

SIMILAR SPECIES The most similar common species is Downy Emerald *Cordulia aenea*, which differs in being slightly smaller and less brilliantly green, with a noticeably downy thorax and a flight pattern that avoids deep shade. It is scattered throughout the area.

Male Brilliant Emerald in flight over a lake.

Four-spotted Chaser

Libellula quadrimaculata

A medium-sized, heavy-bodied, common brown and black dragonfly, Four-spotted Chaser is up to 40–50mm long. Its most distinctive feature is the markings on the wings, which occur in both sexes: the hindwings have a large dark spot at the base, and all four wings have a dark spot in the middle of the leading edge, in addition to the usual pterostigma spot. The females are more orange-brown than the darker brown males. This combination is diagnostic.

FLIGHT PERIOD Early May–late August.

HABITAT AND DISTRIBUTION Common almost throughout northern Europe, except in the far north or on higher mountains, wherever there are suitable clean, well-vegetated water bodies, particularly those with acid water.

SIMILAR SPECIES Females and immatures of the following species, see opposite.

Male Four-spotted Chaser.

Broad-bodied Chaser

Libellula depressa

This is a medium-sized but bulky species that is 40–50mm long and has a noticeably broad abdomen, as its name suggests. Males are very distinctive, with a bright blue, broad body and bold brown patches at the base of each wing. Females and immatures are yellowish-brown, with two broad pale stripes on the thorax. The overall appearance of the female can give rise to confusion with Hornets. This is one of the first dragonflies to appear in spring.

FLIGHT PERIOD Late April–September.

HABITAT AND DISTRIBUTION Common and widespread in England and Wales; rare elsewhere. It occurs most often in small, shallow pools that are frequently bare and open.

SIMILAR SPECIES Blue Chaser *L. fulva* is similar but more slender, and the last three segments of the abdomen are black in males. It is locally common in more vegetated waters, but more southern in distribution and uncommon in the UK. See also the next species.

Male Broad-bodied Chaser perched.

Black-tailed Skimmer
Orthetrum cancellatum

The skimmers are rather similar to the chasers, although they are generally more slender and have completely clear wings that lack dark basal patches. Males of this species have a blue abdomen, with the last few segments being black. Females are yellowish-brown and black. The males are very active fliers and are sometimes known as 'blue arrows'.

FLIGHT PERIOD Late April–September.

HABITAT AND DISTRIBUTION Common and widespread throughout England, Wales and Ireland. Most frequent around large, still or slow-moving water bodies, especially those that are bare and open such as canals and quarry lakes.

SIMILAR SPECIES Keeled Skimmer *O. coerulescens* is more slender with a gradually tapering body; males are pale blue without a black tip; both sexes have distinct yellowish thorax stripes. This is a fast-flying species, although it rarely travels far and frequently settles on vegetation. In most of southern Europe it is widespread and abundant in a variety of slow-flowing waters such as ditches and streams. In northern Europe, including Britain, it is much more likely to be found around bogs and flushes, particularly where there is moving water. It is locally common here where there is suitable habitat, and is largely western within Britain.

Right: Male Black-tailed Skimmer.

Common Darter

Sympetrum striolatum

The darters as a group are reasonably distinctive in the area due to the red coloration of the males (with one exception), the clear wings apart from yellowish patches or coloured veins, and their characteristic behaviour, which involves darting out from and back to a fixed perch, rather in the manner of a Spotted Flycatcher. The most common, though not the most distinctive species, is Common Darter. Males are dull reddish-orange with an almost cylindrical abdomen, black and yellow striped legs, and virtually no colour in the wings. The females of darters are hard to distinguish without detailed examination.

FLIGHT PERIOD June–October.

HABITAT AND DISTRIBUTION Common and widespread throughout almost the whole of Britain in a wide variety of open waters.

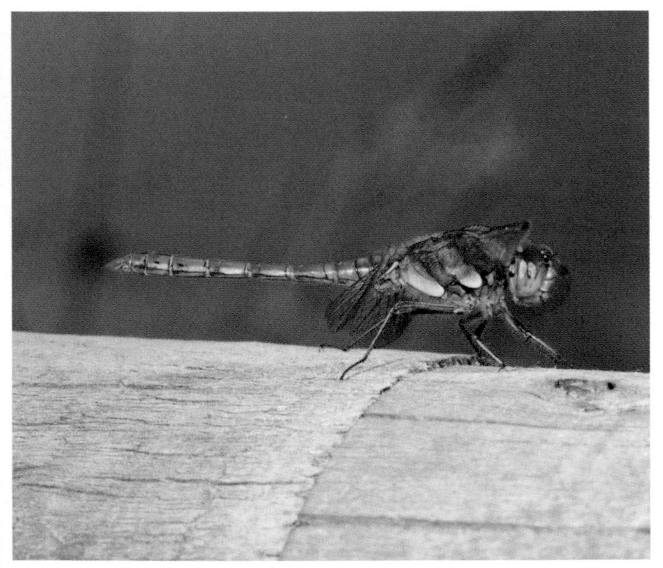

Male Common Darter by a lake.

Ruddy Darters, mating pair.

SIMILAR SPECIES Ten or so darters occur within the area, of which the following are the most frequent. Ruddy Darter *S. sanguineum* is more scarlet in colour, with a conspicuously 'waisted' abdomen and black legs. It is found in similar habitats and has a similar distribution, although it is rather more southern. Red-veined Darter *S. fonscolombii* is best identified by the bright red veins in its wings and the yellowish patches on its hindwings. It is a migrant from the south, often reaching as far north as southern Scotland. Black Darter *S. danae* differs in that males are largely black, although both sexes have a black band on the thorax side, enclosing some yellow blobs. It is widespread, and is most common in heathy and moorland areas with acidic waters.

STONEFLIES, PLECOPTERA

This small group of insects comprises 36 species in the UK. They have soft, flattened bodies and two pairs of wings (though some are flightless), the front pair being smaller than the hind pair. The wings are either held flat against the body when at rest, or wrapped around it to form a cylindrical shape. Many of the most conspicuous species have two long cerci, or tails. None are brightly coloured, and many are rarely noticed. The nymphs are all aquatic, particularly in fast-flowing streams and rivers.

Giant Stonefly
Dinocras cephalotes

Giant Stonefly is one of the largest stoneflies in Britain, with the females reaching about 30mm in length. They are mostly black, with the head having variable amounts of orange-yellow. Males are smaller and have shorter wings. There are two long cerci at the rear. They fly readily in the day. The cast nymph skins may often be found on stones in rivers.

FLIGHT PERIOD
May–August.

HABITAT AND DISTRIBUTION
Widespread but largely confined to western and northern Britain, where there are more oxygenated, fast-flowing streams than elsewhere.

SIMILAR SPECIES *Perlodes microcephala* is almost as large, but has a paler thorax and is more likely to be found in lowland streams.

Mating pair of Giant Stoneflies.

Large Yellow Stonefly
Isoperla grammatica

A medium-sized but quite distinctive stonefly up to 15mm long, this species is known to anglers as 'Large Yellow Sally'. It is mainly yellow with darker markings on the head and thorax. It can hatch in enormous numbers at times, hence its interest to anglers.

FLIGHT PERIOD April–August.

HABITAT AND DISTRIBUTION Common and widespread throughout most of Britain except East Anglia, occurring in stony and gravelly fast-flowing streams, especially over calcareous rocks.

SIMILAR SPECIES The 'Small Yellow Sally' *Chloroperla torrentium* is much smaller at up to 10mm long, and occurs in similar habitats.

Large Yellow Stonefly.

CRICKETS AND GRASSHOPPERS, ORTHOPTERA

This is a distinctive and familiar group of large insects with cylindrical bodies and greatly lengthened back legs. Grasshoppers (Acrididae family) are generally smaller than bush-crickets, with short antennae and no prominent ovipositor in females; they stridulate by rubbing their legs against the wings. Bush-crickets (Tettigoniidae) tend to be larger, are often slightly hunched and have very long antennae, and females have conspicuous ovipositors. They stridulate by rubbing their wings together. True crickets (Gryllidae) are similar, but flatter and broader with a narrow, needle-like ovipositor. Groundhoppers (Tetrigidae) differ from grasshoppers in having an elongated pronotum that extends back over the abdomen, and very short forewings that are reduced to scales.

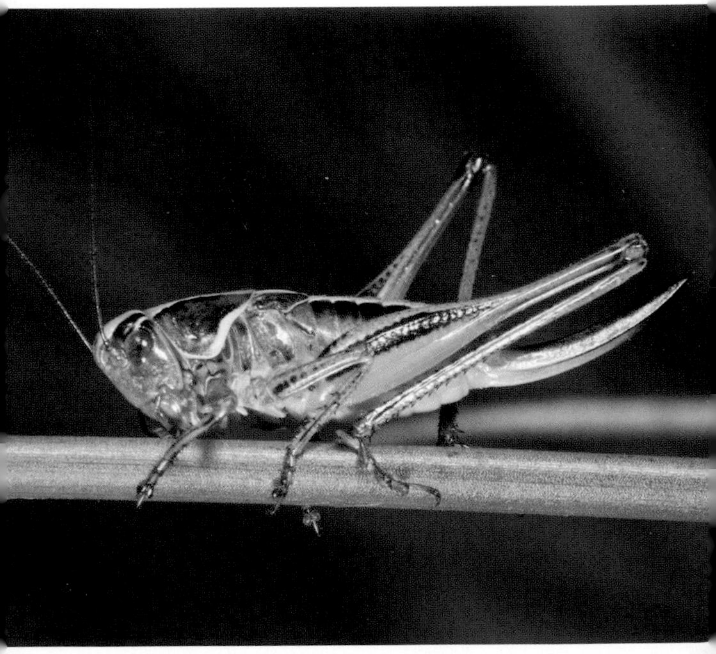

Female Bog Bush-cricket with long ovipositor.

Bush-crickets

Great Green Bush-cricket

Tettigonia viridissima

This species is the largest of the bush-crickets in the region, being up to 55mm in length with antennae at least as long as the body. Overall it is mid-green, and has a brown stripe down the back keel and wings that extend well beyond the body. The female's ovipositor is about 20mm long and slightly downcurved. The song is loud, sounding like a sewing machine and continuing uninterrupted for long periods, especially in the late afternoon and into the night.

FLIGHT PERIOD July–October.

HABITAT AND DISTRIBUTION Local but widespread in warmer parts of northern Europe as far north as southern Scandinavia, but strongly coastal or southern in most areas. Most frequent in sunny places with shrubs or long grass.

SIMILAR SPECIES See Wart-biter.

Female Green Bush-cricket.

Speckled Bush-cricket
Leptophyes punctatissima

A distinctive but inconspicuous small bush-cricket, this species is up to 20mm long with long, slender antennae. The body is green with fine dark speckles all over; the ovipositor is broad and upcurved. The song is virtually inaudible to humans.

FLIGHT PERIOD July–October.

HABITAT AND DISTRIBUTION Common and widespread in southern Britain; rare further north. Adaptable to many bushy habitats, including gardens.

SIMILAR SPECIES Oak Bush-cricket *Meconema thalassinum* is similar in colour, but unspotted and with wings just longer than the body. The ovipositor is relatively long and gently upcurved. The species is widespread in scrub, woodland and gardens, and quite often comes to lights at night.

Male Speckled Bush-cricket on Corn Cockle.

Dark Bush-cricket

Pholidoptera griseoaptera

This is a bulky bush-cricket up to about 20mm long, with long antennae. It is generally dark brown, but always bright greenish-yellow underneath. The wings are short or virtually absent, and the ovipositor is up to 10mm long and upcurved. The call is a short buzz, repeated frequently.

FLIGHT PERIOD July–November.

HABITAT AND DISTRIBUTION Common and widespread in southern Britain, and northwards through Europe to southern Scandinavia, most frequently in warm places with scrub.

SIMILAR SPECIES Bog Bush-cricket *Metrioptera brachyptera* has a similar shape, but is green above and has a white stripe on the back of the pronotum. It is local in southern Britain, in boggy or heathy places. There are several similar species in the region.

Male Dark Bush-cricket basking.

Long-winged Conehead
Conocephalus discolor

This slender, straight bush-cricket is up to 20mm long, with long, fine antennae. The body is green with a brownish stripe on the back, and the wings are brown (in both sexes) and often extend beyond the body. The ovipositor is long, just slightly upcurved and dagger-like. The song sounds like a quiet sewing machine, and is audible (to younger listeners) up to about 2m away.

FLIGHT PERIOD July–October.

HABITAT AND DISTRIBUTION Locally abundant in extreme southern Britain, and slowly spreading north. Favours rough long grassland with plenty of sun.

SIMILAR SPECIES Short-winged Conehead *C. dorsalis* is very similar, but its wings are normally only half the abdomen length, the ovipositor is strongly curved and the song varies in volume. It is common only in southern and eastern Britain, especially in coastal areas.

Female Long-winged Conehead.

Wart-biter

Decticus verrucivorus

A large and bulky bush-cricket, Wart-biter is up to about 40mm long with wings about the length of the body. It is variable in colour, but most commonly green mottled with black, with black eyes. The ovipositor is long (about 20mm in length) and slightly upcurved. The unusual name derives from an early practice, mainly in Scandinavia, of using it to bite warts on human skin.

FLIGHT PERIOD Adults visible July–October.

HABITAT AND DISTRIBUTION Most common in shortish calcareous grassland, although it also occurs on heathland and in mountain pastures. Widespread and locally common in Europe as far north as Scandinavia; very rare in the UK, in just a few sites in the extreme south.

SIMILAR SPECIES Most similar to Great Green Bush-cricket (see p. 37), differing in shorter wings, more mottled body and black eyes.

A female Wart-biter.

True crickets

Wood-cricket
Nemobius sylvestris

This is an inconspicuous little insect visually, but it is often detected by its attractive mellifluous purring song. It is frequently abundant where it does occur. The body is brown and shiny, up to 10mm long, with a yellowish pronotum, two long 'tails' at the rear and two long antennae. The wings cover less than half of the abdomen, and the insect is flightless. The sexes are similar, except for the female's thin ovipositor.

FLIGHT PERIOD May be seen all year, most frequently April–November.

HABITAT AND DISTRIBUTION Local in southern England; more widespread in central Europe as far north as Germany in and around deciduous woodland.

SIMILAR SPECIES House Cricket *Acheta domestica* is similar in shape, but twice as large, paler, and fully winged and able to fly. It is an introduced species that is locally established in warm buildings.

Female Wood-cricket in woodland.

Field Cricket

Gryllus campestris

Field Cricket is an attractive and conspicuous insect with a shiny black body and sculpted blackish-brown wings that have a yellow band next to the thorax and reach the length of the abdomen. Superficially it resembles ground beetles (see p. 152), and is often confused with them. Both sexes have two thin 'tails', and the female has an ovipositor. Field Crickets live in burrows. The male sits at the mouth of the burrow, from which it chirps endlessly, retreating if danger threatens, although males are not infrequently found away from the burrows.

FLIGHT PERIOD May–September.

HABITAT AND DISTRIBUTION Very rare in southern Britain and northwards through Europe to southern Scandinavia, most frequently occurring in warm, grassy places where it may be abundant and readily audible. Most frequent in southern Europe.

SIMILAR SPECIES None.

Female Field Cricket.

Grasshoppers

Common Field Grasshopper

Chorthippus brunneus

This is a medium-sized grasshopper that is generally pale brownish in colour, about 20mm (males) and 25mm (females) long, with noticeably long wings that extend beyond the abdomen. The underside of the body is hairy, and the tip of the abdomen is often reddish. The song of the male is a short, buzz-like chirp, repeated up to 10 times at about 1–3 second intervals, often with other nearby males joining in.

FLIGHT PERIOD June–October.

HABITAT AND DISTRIBUTION Widespread and common throughout the British Isles in suitable habitats. Most common in short grass and other vegetation in warm, dry habitats.

Common Field Grasshopper.

Meadow Grasshopper – note the short wings.

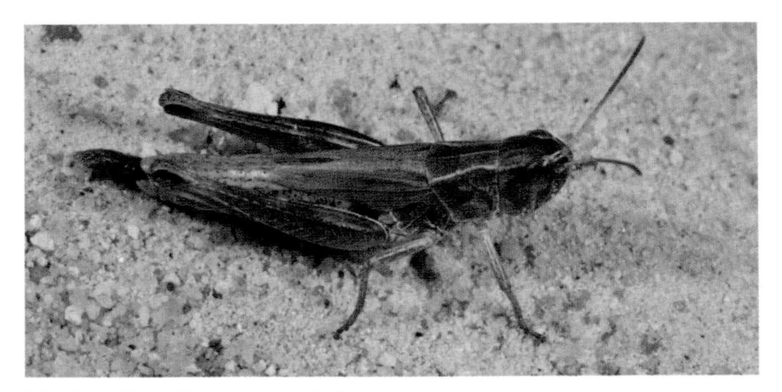

Meadow Grasshopper, purple form.

SIMILAR SPECIES Meadow Grasshopper *C. parallelus* is similar, but usually more green and with very short wings. Its call is a brief burst of scratchy calls lasting 1–2 seconds in all. It is common throughout the area, usually in lusher, taller and often damper habitats. Common Green Grasshopper *Omocestus viridulus* is generally similar, usually all green, and never with red on the abdomen. The wings are long, but shorter than the abdomen, and the species is able to fly. The call is a continuous rapid ticking, which rises in volume for 8–10 seconds, then levels out for about the same length of time. It is common and widespread in drier grasslands, and often the first species to mature, in June. There are several other similar species in the region, which need detailed examination for certain identification.

Large Marsh Grasshopper
Stethophyma grossum

This large and conspicuous grasshopper is up to 25mm (males) and 40mm (females) long. It is variably marked with green, brown and black, but has a striking red patch on the hind legs and its wings are distinctly longer than the body. Both sexes can fly well (for grasshoppers), becoming very noticeable and often travelling 30–40m before they settle. The call of the male is a series of short, loud ticks.

 **FLIGHT PERIOD** July–October.

HABITAT AND DISTRIBUTION Local in southern Britain and western Ireland; absent elsewhere in the UK. In mainland Europe it is found in bogs, fens and wet meadows, although it is confined solely to large bogs in Britain.

SIMILAR SPECIES None in UK.

Male Large Marsh Grasshopper.

*Blue-winged Grasshopper
Oedipoda caerulescens

A number of European grasshoppers, particularly in the south, have developed 'flash colours' as a form of camouflage. When they fly they show bright red or blue hindwings, which disappear when they settle, leaving a predator looking for a red or blue insect. This species is greyish-brown to dark brown, often striped darker, up to 30mm long (female) and has blue hindwings edged with black.

FLIGHT PERIOD July–October.

HABITAT AND DISTRIBUTION Absent from Britain, but locally common in mainland Europe as far north as southern Sweden and Norway, in rough, warm, grassy places, especially in hilly areas. Declining in the north of its range.

SIMILAR SPECIES *Red-winged Grasshopper *Oedipoda germanica* is similar, but red-flashing and more southern.

Blue-winged Grasshopper showing part of blue hindwing.

Mottled Grasshopper

Myrmeleotettix maculatus

The female of this small and very variable grasshopper is up to 18mm long, while the male is smaller. As the name suggests, it is usually mottled but may be any combination of green, grey, reddish-brown and white, although the wings are never green. The wings of the male reach to the end of the abdomen, while those of the female are slightly shorter. The male's antennae are distinctly clubbed and usually bent outwards (an unusual feature among grasshoppers); those of the female are much less so. The call sounds like the winding of a clock.

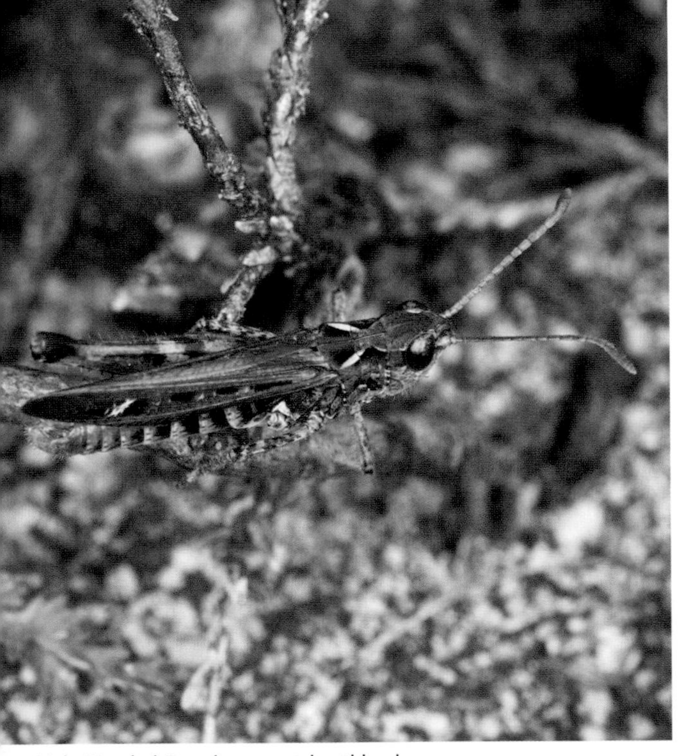

Male Mottled Grasshopper on heathland.

FLIGHT PERIOD June–October.

HABITAT AND DISTRIBUTION Widespread and quite common throughout the British Isles in warm, dry habitats, although easily overlooked because of its small size.

SIMILAR SPECIES In Rufous Grasshopper *Gomphocerippus rufus* the male's antennae are also thickened at the tip, but they are white-tipped and not bent. It is a slightly larger species, up to 23mm long and rather plain in colour. It is uncommon and southern in Britain in warm places on limestone, and more widespread and common in Europe.

Male Rufous Grasshopper – note white-tipped antennae.

Groundhoppers

Common Groundhopper

Tetrix undulata

Groundhoppers are flightless, but are able to jump well. They feed mainly on mosses and algae, requiring patches of damp bare ground in their territory, as well as some cover. They are also occasionally found in greenhouses. This small species is about 10mm long, mottled with brown, blackish or reddish. The elongated pronotum is strongly keeled and does not reach beyond the tip of the abdomen, and the hindwings are shorter than the pronotum. It is flightless and has no call.

FLIGHT PERIOD Active all year, most noticeable April–October.

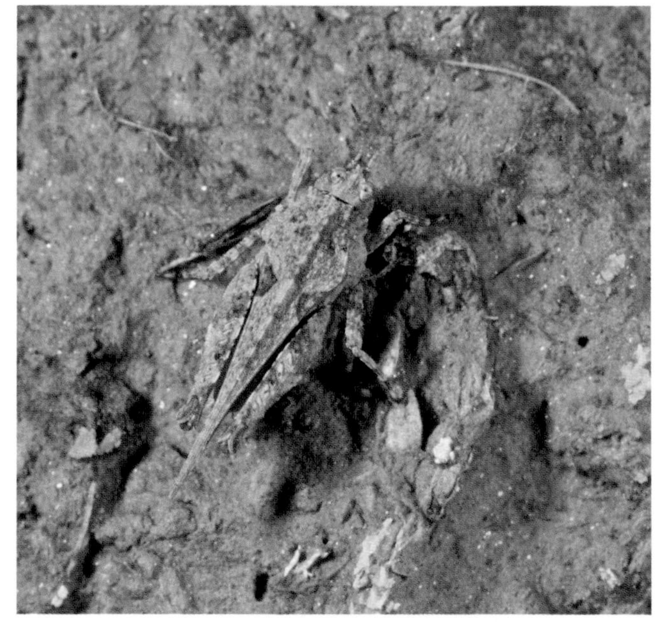

Cepero's Groundhopper on mud.

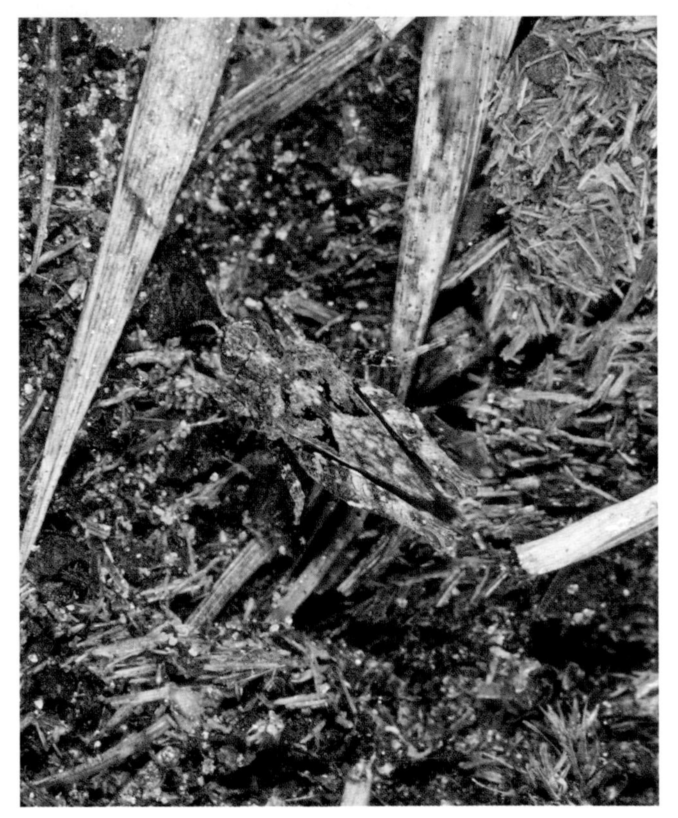

Common Groundhopper on heathland.

HABITAT AND DISTRIBUTION Common in southern Britain and Ireland, and widespread in northern mainland Europe. Occurs in a variety of habitats where there is partially sunny bare ground.

SIMILAR SPECIES Slender Grasshopper *T. subulata* has a pronotum that is longer than the abdomen, and longer hindwings. It is widespread in similar although often damper habitats. The rare Cepero's Groundhopper *T. ceperoi* is very similar to Slender Grasshopper, and is best separated by the distance between the eyes – it is twice the width of the eye in Slender, 1.5 times the width in Cepero's.

COCKROACHES AND MANTISES, DICTYOPTERA

These insects characteristically have long antennae and bristly legs. Of 3,500 cockroach species worldwide, three are found in Britain.

Common Cockroach

Blatta orientalis

The cockroaches are medium to large insects with flattened bodies, long antennae and spiky legs. This species, sometimes known as Black Beetle, is dark shiny brown or black and 20–30mm long. Females have vestigial wings and males are winged, but neither sex flies.

FLIGHT PERIOD Mainly summer, depending on the warmth of the habitat.

HABITAT AND DISTRIBUTION Introduced from Africa and now widespread in heated environments, but rarely found outdoors.

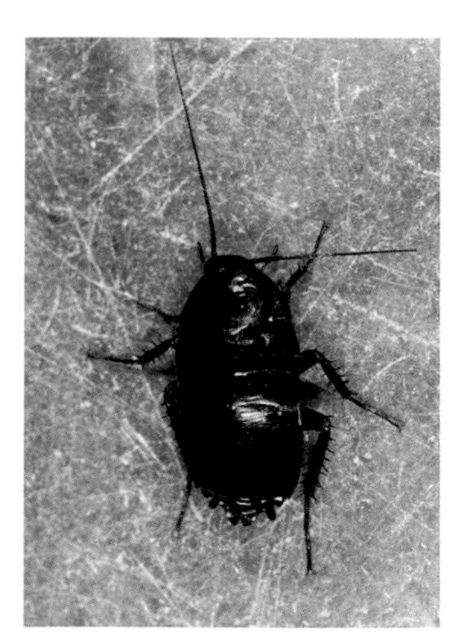

Common Cockroach.

SIMILAR SPECIES There are several native species, which are always much smaller, up to 10mm long, with fully formed wings and able to fly well. Tawny Cockroach *Ectobius pallidus* is one of several similar species found in southern Britain and more widely in northern Europe, in open habitats and woodland clearings.

*Praying Mantis

Mantis religiosa

Praying Mantis is a highly distinctive, very large insect, up to 60mm long, with a green or (rarely) brown body and wings. It has a small, mobile head, wings that are as long as the body and enlarged powerful front legs. Females are most frequently seen, partly because males are smaller, but also because they are often eaten by the females after mating. The egg masses consist of distinctive large, frothy, solidified masses, which are deposited on wood, stones and vegetation.

FLIGHT PERIOD July–October.

HABITAT AND DISTRIBUTION Found in warm, grassy and scrubby places. Most common in southern Europe, but extends almost throughout France and to scattered sites in Germany. Absent from Britain.

SIMILAR SPECIES None.

A female Praying Mantis.

EARWIGS, DERMAPTERA

These elongated insects are nocturnal ground-living scavengers that eat both plant and animal matter.

Common Earwig
Forficula auricularia

This familiar and abundant, shiny brown insect is just over 10mm long, with long antennae and tweezer-like pincers at the rear end, which are strongly curved in males and roughly straight in females. Although the insects appear wingless, the hindwings are concealed under elytra, but the species rarely flies. The females guard their eggs and sometimes the developing young, which are like miniature adults.

FLIGHT PERIOD Adults found all year round, although they are mainly active in warmer periods.

HABITAT AND DISTRIBUTION Widespread and common just about everywhere, wherever there is food and shelter, especially in gardens.

SIMILAR SPECIES There are several species, although none is common.

Adult Common Earwig.

TRUE BUGS, HEMIPTERA

This is a huge order of insects, with at least 8,000 species in Europe alone. They occur in a great range of forms and sizes, and are linked primarily by having a syringe-like, piercing beak, the rostrum, with which they suck juices. There are two main suborders: the Heteroptera, such as shield bugs, in which the wings are divided into a horny basal part and a membranous tip, normally held flat over the body at rest; and the Homoptera, which include aphids and leaf hoppers, with uniform wings that are either membranous or leathery, and usually held roof-like over the body.

A Green Shield Bug nymph.

Heteroptera

Gorse Shield Bug
Piezodorus lituratus

This is a typical shieldbug with the two-part wings characteristic of the heteropteran bugs. It is about 13mm long, narrowly oval in shape and highly variable in colour. Overwintered adults appear bright green in spring, whereas the summer generation has red on the wing-cases and pronotum. They feed mainly on gorse and less frequently on other shrubby legumes.

FLIGHT PERIOD Adults found all year round, although they are usually dormant in winter.

HABITAT AND DISTRIBUTION Widespread and common throughout mainland Britain, wherever gorse occurs.

SIMILAR SPECIES Hawthorn Shield Bug *Acanthosoma haemorrhoidale* is similar, but has more pointed 'shoulders' and a tapered abdomen. It is common on hawthorn and other shrubs.

Gorse Shield Bugs on Gorse.

Forest Bug or Red-legged Shieldbug
Pentatoma rufipes

A large and conspicuous shield bug up to 15mm long, this species is predominantly shiny brown, with orange-red legs and a small cream to orange area at the tip of the scutellum. The sides of the pronotum are extended into projecting and slightly hooked 'shoulders'. The adults are omnivorous, acting as predators feeding on other invertebrates as well as vegetable matter.

FLIGHT PERIOD June–November, although occasionally also seen in spring.

HABITAT AND DISTRIBUTION Widespread and common throughout the area, except Ireland, in woods, parks, gardens and orchards.

SIMILAR SPECIES *Troilus luridus* is similar but has more rounded shoulders and a yellow band on each antenna. It is widespread in woodlands.

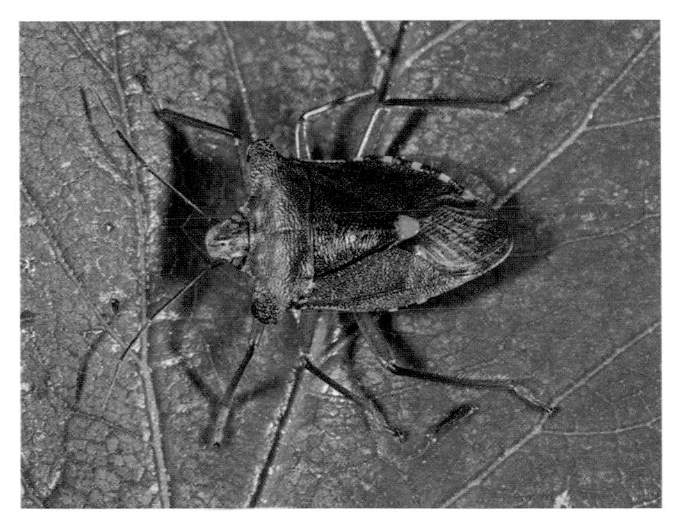

An adult Forest Bug.

Brassica Bug or Crucifer Shield Bug
Eurydema oleracea

This small and variable but distinctive bug is oval in shape and up to 7 mm long. It is essentially shiny dark green or bluish-black, with markings that may be red, cream, white or orange, generally becoming paler in winter.

FLIGHT PERIOD Adults occur all year round, but mainly active April–October.

HABITAT AND DISTRIBUTION Locally common in southern England, and more widespread in northern mainland Europe except in the far north. Feeds on wild brassicas such as Hedge Mustard, and occasionally on crops.

SIMILAR SPECIES Ornate Shieldbug *E. ornata* is similar, but larger and more boldly coloured. It is a new arrival in Britain, and found locally throughout northern mainland Europe as far north as Denmark.

Paired Brassica Bugs in their black and white form.

Green Shield Bug

Palomena prasina

This is probably the most common and distinctive of the shield bugs, large enough to be noticeable (up to 15mm long), and almost entirely green in colour except for the dark, membranous wing-tips and fine black dots all over the surface. Hibernating adults darken and turn bronze-brown in autumn, then green again when they emerge in spring. They feed on a wide variety of deciduous trees and shrubs.

FLIGHT PERIOD Adults occur all year round, but hibernate, and are mainly active April–October, depending on the weather.

HABITAT AND DISTRIBUTION Widespread throughout Britain and much of continental Europe (though rarer in the north), anywhere where there are broadleaved trees and shrubs.

SIMILAR SPECIES None when it is in its green phase.

Adult Green Shield Bug in summer.

Dock Bug
Coreus marginatus

This large member of the squash bug group is roughly oval to diamond shaped in outline, up to 15mm long and generally dull brown in colour. The antennae are long, four-segmented and yellowish-orange, with two tiny, horn-like projections between them.

FLIGHT PERIOD Adults occur all year round, but hibernate, and are mainly active May–October, with noticeable concentrations before hibernation in autumn.

HABITAT AND DISTRIBUTION Feeds mainly on the seeds and foliage of dock, supplemented by other fruits, and can occur wherever docks are found. Mainly southern in Britain, but widespread in continental Europe north to southern Scandinavia.

SIMILAR SPECIES There are several similar but smaller species, such as *Syromastes rhombeus*, which has shorter wings. Mainly southern, and found in heathy areas.

Mating Dock Bugs.

*Italian Striped-bug or Minstrel Bug

Graphosoma italica

This strikingly marked and conspicuous bug is about 10–12mm long. Its whole body, from the head to the tips of the wing-cases, is boldly vertically striped with red and black. This strong warning colouration warns predators of its bitter taste, and allows it to openly spend its life without seeking cover – which makes it highly likely to be seen by us. The bugs cluster on umbellifers and other herbaceous plants in large numbers.

FLIGHT PERIOD June–October.

HABITAT AND DISTRIBUTION Found in warm, flowery places, from central/western Europe southwards; absent from Britain.

SIMILAR SPECIES None in the area, although there are similar species in southern Europe.

An Italian Striped-bug.

Firebug

Pyrrhocoris apterus

Firebug is an attractive and sociable little insect that is boldly black and red in colour, with a large, circular black spot on each wing. Firebugs are normally short-winged (and flightless), with the lower third of the abdomen remaining exposed, but a few individuals have longer wings. The head is all black.

FLIGHT PERIOD Adults occur all year round, but hibernate; they emerge in spring, when they may gather in sizeable and active groups.

HABITAT AND DISTRIBUTION Very rare, in south-west England only, but quite common on the Continent in warmer places, as far north as Sweden and Norway. Most commonly found in warm, sheltered natural habitats such as woodland clearings, wood margins and south-facing coasts.

SIMILAR SPECIES There is another bug, *Corizus hyoscyami*, which looks similar but is always long-winged, has a red and black head and does not occur in dense groups. It is uncommon and mainly coastal.

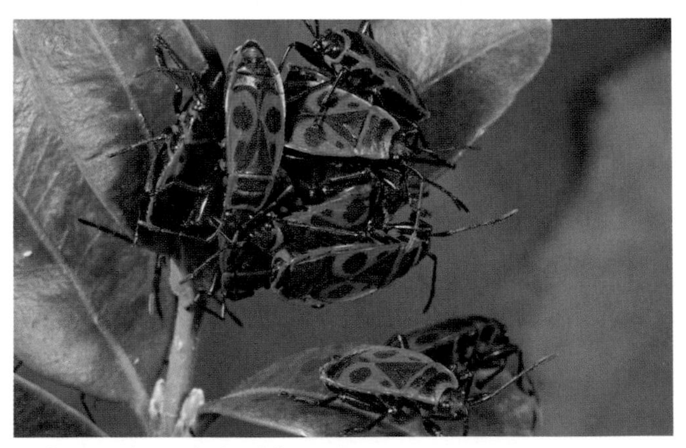

Firebugs gathering before hibernation.

Common Green Capsid

Lygocoris pabulinus

This medium-small insect measures up to only about 7mm in length, but is quite conspicuous because of its bright green colour and its abundance. The wings are green except for the brown membranous wing-tips, there is a distinct pale collar at the front of the pronotum and there are fine pale yellowish spines on the legs.

FLIGHT PERIOD Mainly seen in early June–October.

HABITAT AND DISTRIBUTION Abundant throughout the British Isles and much of northern Europe. It needs a combination of woody and herbaceous plants for the various parts of its life-cycle, so is most common on woodland edges, scrub, meadows and hedges, also becoming a pest in orchards at times.

SIMILAR SPECIES There are many similar plant-feeding bugs, although the combination described above is distinctive.

Common Green Capsid on Knapweed.

Common Pondskater

Gerris lacustris

A common and widespread insect, this species measures up to just over 10mm long, and has a narrow, cylindrical body that is brown or grey-brown in colour, and a strong, downcurved beak. It has a conspicuous habit of skating across the water's surface, casting shadows where its feet dent the surface film of the water, and finding dead or stranded insects to eat. Some individuals are winged, others wingless.

FLIGHT PERIOD Adult all year, but hibernates through the coldest periods.

HABITAT AND DISTRIBUTION Abundant throughout the British Isles on still or slow-moving waters.

SIMILAR SPECIES There are several other pond skaters. Water Cricket (see p. 68) is similar, but shorter and broader, less active and more likely to be in the shade.

Common Pondskater adult.

River Skater

Aquarias najas

This is a distinctly bigger species than Common Pondskater, up to 18mm long and occasionally larger. It is similar in colour and appearance, but looks much longer and narrower, and the final segment of the abdomen has a short, pointed projection on each side. It is flightless and spends its time on the flowing water's surface, constantly rowing to maintain position against the flow. After mating in spring the female submerges to lay eggs on the surfaces of stones underwater.

FLIGHT PERIOD Adults occur all year round and may be seen at any time.

HABITAT AND DISTRIBUTION Most frequently found on well-oxygenated rivers, and occasionally occurs on lakes. It is mainly western in its distribution, and rare in Scotland.

SIMILAR SPECIES Common Pondskater (see opposite).

Adult River Skater.

Water Scorpion

Nepa cinerea

Water Scorpion is a curious and very distinctive large insect, with a long, flattened body and a long, thin tail resembling the sting of a scorpion, but actually used as a breathing tube. It is up to 22mm long excluding the tail. Water Scorpions are slow moving and may remain partly submerged for long periods, breathing through the tail. They hunt by remaining quite still, then catching unsuspecting invertebrates.

FLIGHT PERIOD Adult all year round, but dormant through cold periods.

HABITAT AND DISTRIBUTION Widespread throughout the British Isles and much of northern mainland Europe, although rarely common, in shallow, well-vegetated still waters.

SIMILAR SPECIES Nothing is very similar, but see Water Stick Insect, opposite.

A Water Scorpion submerged.

Water Stick Insect

Ranatra linearis

Although closely related to Water Scorpion, this species could not be confused with it, but it is another highly distinctive aquatic bug, with a long, thin, stick-like body and tail up to 40mm long. In some respects it resembles, although is not related to, a Praying Mantis (see p. 53) in both structure and habits. It is surprisingly inconspicuous for such a large insect, as it spends much of its time motionless below water waiting for prey to come close, although it does also fly in warm weather.

FLIGHT PERIOD Adult all year round, but dormant through cold periods.

HABITAT AND DISTRIBUTION Found in southern England and Wales only, in deep, well-vegetated ponds.

SIMILAR SPECIES Water-measurer *Hydrometra stagnorum* is much smaller, up to 12mm in length, and hunts by walking slowly on the water's surface.

Water Stick Insect.

Water Cricket

Velia caprai

This is a curious and distinctive little aquatic bug, resembling a less streamlined Common Pondskater and slightly reminiscent of a cricket. Adults are up to 10mm long, but may be conspicuous due to their habit of occurring in large groups. They are dark brown, but in good light can be seen to be edged with orange or yellow. They scavenge on the water's surface for dead and dying invertebrates. Although generally quite slow moving, they can travel more quickly if necessary by ejecting saliva onto the water's surface, thus lowering the surface tension.

FLIGHT PERIOD Most likely to be seen February–October.

HABITAT AND DISTRIBUTION Widespread throughout the British Isles and much of northern mainland Europe, on the surfaces of shady streams and ponds, especially in upland areas.

SIMILAR SPECIES Common Pondskater (see p. 64).

Two Water Crickets on a river surface.

Common Backswimmer or Water Boatman

Notonecta glauca

This medium-large aquatic bug has the distinctive and endearing habit of swimming on its back, with a large silvery air bubble attached to its belly, and powerful back legs which it uses like oars. It is about 15mm long and greyish-brown, although it appears silvery because of the bubble, and is highly predatory on other invertebrates or even small tadpoles.

FLIGHT PERIOD Adults occur all year round, but are dormant through cold periods.

HABITAT AND DISTRIBUTION Widespread and quite common throughout the British Isles and much of northern Europe, in still or slow-moving waters.

SIMILAR SPECIES Lesser Water Boatman *Corixa punctata* is smaller and swims the right way up, in similar habitats. It is mostly southern in Britain, and widespread in continental Europe north to southern Scandinavia.

Common Backswimmer, swimming upside-down.

Homoptera

Horned Treehopper

Centrotus cornutus

A small though distinctive little insect up to 6–7mm long, this species is dark brown in colour with a hunched look due to the keeled pronotum. The pronotum is leathery and about as long as the abdomen, and the brown-tinted wings are slightly longer. On each 'shoulder' of the pronotum there is a pointed projection. The insects are most commonly seen attached to the stems of herbaceous plants such as Marsh Thistle, from which they suck the sap.

FLIGHT PERIOD Early May–August.

HABITAT AND DISTRIBUTION Widespread and quite common throughout most of Britain, but absent from Ireland, in sheltered flowery places such as woodland rides and glades.

SIMILAR SPECIES A smaller species, *Gargara genistae*, has no horns and a shorter pronotum, and occurs mainly on broom and greenweeds.

Horned Treehopper on Marsh Thistle.

Common Froghopper or Cuckoo-spit

Philaenus spumarius

The early nymphal stages of this species are very familiar as the white, frothy masses found on plants in spring, known generally as 'cuckoo-spit' because they often appear at the same time as Cuckoos first arrive. The much less familiar adults emerge from this froth, maturing into inconspicuous brownish-patterned hoppers measuring 5–7mm in length, which feed on various herbaceous plants.

FLIGHT PERIOD Cuckoo-spit usually seen April–May, adults June–September.

HABITAT AND DISTRIBUTION Widespread and common throughout the British Isles and much of northern mainland Europe, except in the coldest parts.

SIMILAR SPECIES *Aphrophora alni* is larger, has a distinct keel on the pronotum and is found mainly on trees. The nymphs do not hide in froth.

Common Froghopper nymphs in 'cuckoo spit'.

Red-and-black Froghopper

Cercopis vulnerata

Though small (9–12mm long), this froghopper is highly conspicuous and distinctive (it is actually one of our largest homopterans) due to its boldly mottled red and black colouring, consisting of large, squarish red patches on a black background. The nymphal stages live underground in hardened froth, emerging as brightly coloured adults that may often be seen in large numbers in favoured sites.

FLIGHT PERIOD April–July; particularly noticeable in spring.

HABITAT AND DISTRIBUTION Most common in warm, bushy, sheltered places with long vegetation, such as woodland margins and clearings. In the British Isles it is largely confined to England and Wales, and is rare in southern Scotland.

SIMILAR SPECIES None.

Two Red-and-black Froghoppers.

Green Leafhopper
Cicadella viridis

This is a distinctive and abundant little leafhopper, up to 9mm long, with strongly green or bluish wings that make it quite conspicuous. The leafhoppers can be distinguished from the froghoppers by their very spiny back legs. In favoured places hundreds can hop away in front of you as you walk through the grass.

FLIGHT PERIOD July–October, peaking around August.

HABITAT AND DISTRIBUTION Common in a variety of open, sunny habitats, particularly in wet meadows, fens and marshes, feeding on grasses. Occurs almost everywhere except in the coldest and driest places.

SIMILAR SPECIES Rhododendron Leafhopper *Graphocephala fennahi* is similar in shape and general colour, but is boldly marked with V-shaped red stripes. It was introduced from North America, but is now quite common in parts of southern Britain and mainland Europe as far north as the Netherlands, on rhododendrons.

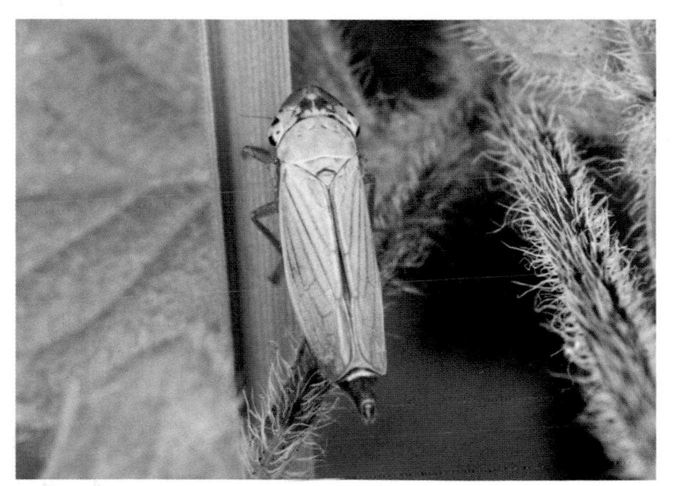

Female Green Leafhopper.

Aphids

Aphids are small, ovoid or pear-shaped, sap-sucking homopteran bugs up to 7mm long, although usually less. They belong to the Sternorrhyncha suborder and there are about 500 species in the UK alone (and many more elsewhere in Europe), in several different families. Although they are generally easy enough to identify as a group, they are extremely difficult to identify to species without a microscope.

Aphids have fascinating life-cycles. A common pattern is for overwintering eggs to hatch in spring into wingless females only, which rapidly give birth, parthenogenetically, to large numbers of new aphids. Most commonly they are born as live young at this stage. As the season progresses more winged forms are produced, and eventually males and females, so that sexual reproduction can take place to allow the laying of the overwintering eggs. The

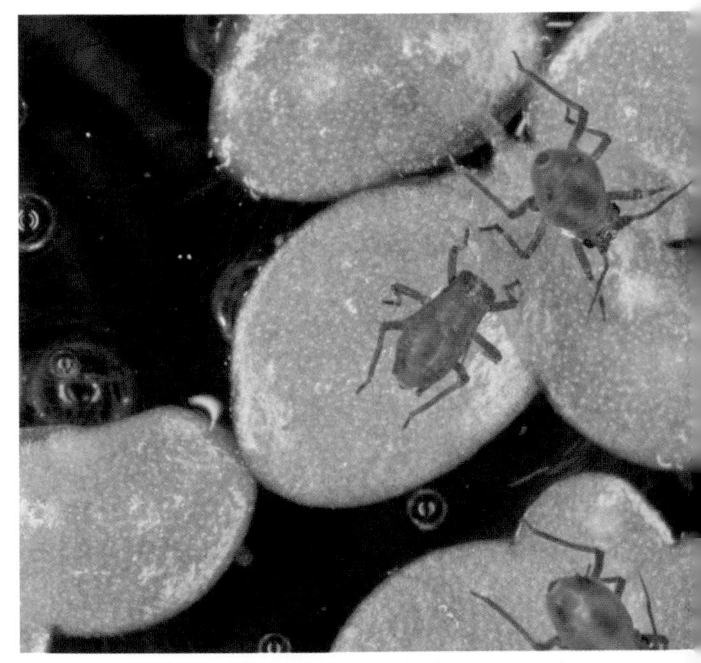

Water-lily Aphids, at different stages, on floating duckweed.

reproduction rate is huge, and some species are known to have up to 40 generations in a year.

A number of aphid species and groups are familiar pests of crops and garden plants, resulting in common names such as 'blackfly' and 'greenfly'. Many feed on different plants at different times of the year, frequently changing between woody and herbaceous plants for their alternate hosts. Some species transmit diseases of crops, particularly viruses. In a number of species there is an interesting relationship between ants and aphids, in which the ants 'farm' the aphids, looking after them and milking them for their honeydew.

The species illustrated, Water-lily Aphid *Rhopalosiphum nymphaeae*, is common on duckweed, water-lilies and other aquatic plants.

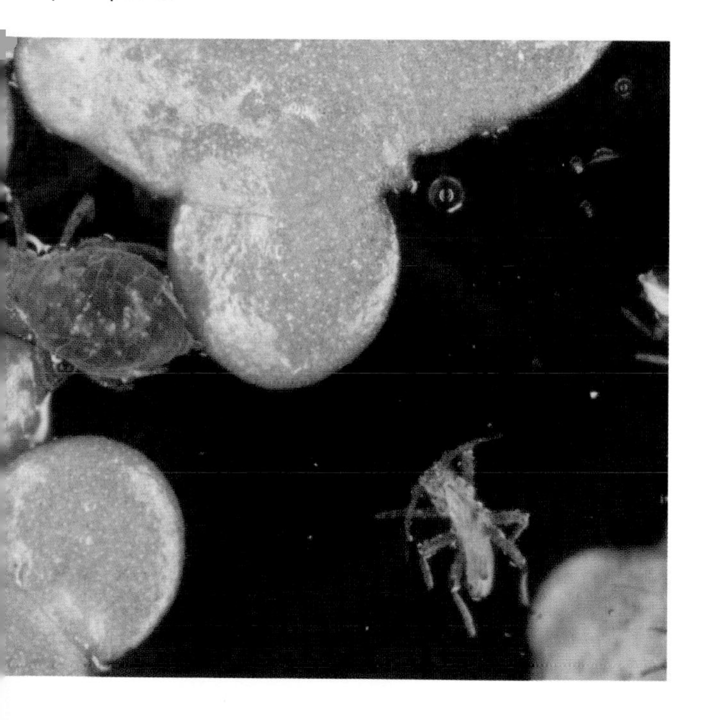

Mealy Plum Aphid
Hyalopterus pruni

This is a typical common aphid, with a complex life-cycle. Its eggs are laid in autumn on the leaves of plum trees, and they hatch in spring to build up large colonies on the undersides of plum leaves, causing some damage to them. By June–July winged forms are produced, which migrate to the leaves of reeds (as shown below) and other waterside grasses. Some adults then return to plum trees, while others do not, and some never leave the trees. Individuals are waxy pale green or pink, and up to 2.6mm long.

FLIGHT PERIOD April–October.

HABITAT AND DISTRIBUTION Widespread and common.

SIMILAR SPECIES Hard to distinguish from many other aphids, although the food plant choice helps.

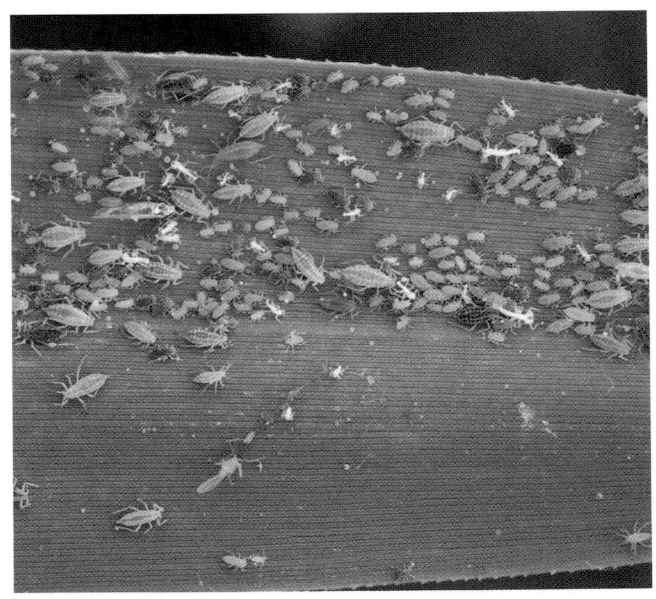

Mealy Plum Aphids on reed.

ANT-LIONS, LACEWINGS AND ALLIES, NEUROPTERA

This is a rather small order (in Britain) of insects with two pairs of roughly equal membranous wings, and full metamorphosis in their life-cycles. Most species are predators on other invertebrates.

*Owl-fly or Ascalaphid
Libelloides coccajus

The owl-flies are among the most conspicuous and attractive members of the Neuroptera, although none occurs in Britain. They are large insects with clear yellow and black wings up to 70mm across, and long, club-tipped antennae. They are aerial predators, resembling dragonflies in the way that they chase prey on the wing. There are several similar species in western Europe, extending north as far as south-west Germany and northern France.

FLIGHT PERIOD Late April–early August.

HABITAT AND DISTRIBUTION Occurs in dry, flowery grassland and woodland clearings.

SIMILAR SPECIES Readily distinguished from dragonflies by the long, clubbed antennae.

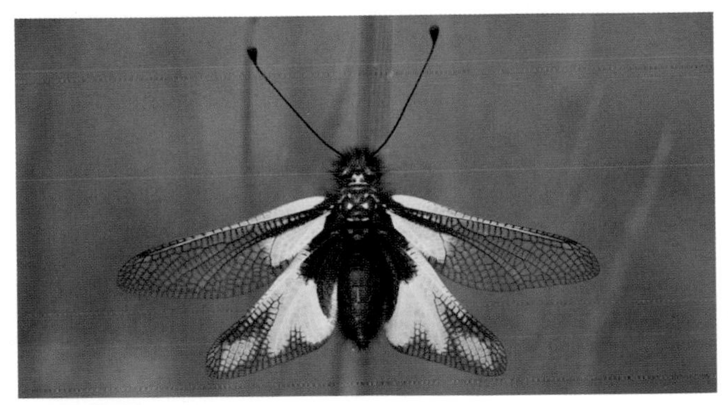

An Owl-fly, basking.

Giant Lacewing
Osmylus fulvicephalus

This is Britain's largest lacewing, readily distinguished from other lacewings by its large size (up to 25mm long) and clear brownish, dark-spotted wings. As in other lacewings, the wings are held like a tent over the body when at rest. The larvae are predators, hunting mainly among damp moss and streamside debris. This is a mainly nocturnal species that rests during the day.

FLIGHT PERIOD Late April–early August.

HABITAT AND DISTRIBUTION Almost always occurs in damp, shady places along streams and small rivers, frequently hiding under bridges and parapets during the day. Locally common in England and Wales, but virtually absent from Scotland.

SIMILAR SPECIES The brown lacewings are similar, but less than half its size. The brownest of the green lacewings, such as *Nothochrysa* spp., are rather smaller and have unspotted wings.

Giant Lacewing on blackberry blossom.

Green Lacewing

Chrysopa carnea

There are several rather similar species of green lacewing, and there may prove to be even more because the group is undergoing taxonomic revision. This species is all green, lacking any black spots or veins, about 15mm long, very delicate and unusual in that it turns pale fleshy-brown before hibernation. It is the only green lacewing that hibernates. Both adults and larvae are predators, mainly on aphids. The eggs are unusual and distinctive – white and laid on long, thin stalks under leaves.

FLIGHT PERIOD All year, but hibernates during coldest periods.

HABITAT AND DISTRIBUTION Widespread and common in woodland, gardens and many other habitats.

SIMILAR SPECIES *C. perla* is more bluish and has extensive black markings around the head. It is found in similar habitats, but is absent from the north and Ireland.

Common Green Lacewing on Moss.

ALDER FLIES, MEGALOPTERA

This small order contains about 300 species worldwide, all with large wings, short lives, and crepuscular or nocturnal habits.

Alder Fly

Sialis lutaria

This is a medium-sized, slow-flying insect that is about 12–15mm long (although males are slightly smaller). It has translucent smoky wings that are held tent-wise over the body when at rest. The wings are hairless and have noticeably thick, dark, conspicuous veins. This alder fly is not a true fly, but one of the few UK members of the Megaloptera order. The adults only live for a week or two, and rarely feed.

FLIGHT PERIOD May–July.

HABITAT AND DISTRIBUTION Locally common almost throughout the area, close to still or slow-flowing waters with plenty of mud, in which the larvae feed. They can survive some degree of pollution.

SIMILAR SPECIES There are two other *Sialis* species, distinguishable only microscopically; caddis flies are similar, but have hairy wings and thinner veins.

Adult Alder Fly.

SCORPION FLIES, MECOPTERA

Scorpion flies are curious and distinctive insects. They derive their name from the female's upturned tail, which looks very like the sting of a scorpion, although it is quite harmless to humans. The male's abdomen is straight and tapering.

Scorpion Fly

Panorpa communis

Both sexes of this species are about 12–15mm long, with membranous wings covered with dark, squarish spots. They have large eyes, and the head has a prominent beak. The insects scavenge for food, including on spiders' webs.

FLIGHT PERIOD May–August.

HABITAT AND DISTRIBUTION Locally common almost throughout the area, in damp or shady, well-vegetated places such as woodland margins, nettle patches and so on.

SIMILAR SPECIES There are two other species in the UK (and many more in continental Europe), which differ in small variable details only and need close examination for accurate identification.

Male Scorpion Fly.

BUTTERFLIES AND MOTHS, LEPIDOPTERA

Butterflies are covered fully in another guide in this series and are therefore only mentioned in passing here, to describe the range of colours and forms. Butterflies and moths are classified together as the Lepidoptera, all of which have distinctive overlapping scales on their wings and, although popularly separated into these two groups, they are actually more closely intertwined taxonomically.

Generally speaking, butterflies (suborder Rhopalocera) in northern Europe are day flying and have thin antennae with distinct club-shaped tips. Moths (suborder Heterocera) are mostly night flying and have variable antennae, but none that are thin with a club at the tip. Most (but not quite all) moths have a little hook – the frenulum – that attaches the two wings of each pair together, whereas no butterflies have this feature. In Britain there are about 60 species of butterfly, which can be divided into seven reasonably distinctive groups, described below.

The moths are a very large group of insects with at least 2,500 species in the UK. The large number of species demands detailed coverage, and there are a number of excellent field guides available (see further reading, p. 188). Some of the most obvious moth groups are briefly described below.

Butterflies

Skippers Small, rather moth-like butterflies, generally overall brownish or orange in colour. Several of the eight UK species hold their wings half erect when settled, unlike any other UK species. The antennae are held widely apart when at rest.

Female Large Skipper.

Common Swallowtail, Norfolk.

Swallowtails In Britain there is only one resident species from this spectacular family of large black, yellow and red butterflies with pronounced 'tails' on the hindwings. They are unlikely to be confused with any other native species.

Whites and yellows Familiar yellow or white butterflies with strong, fluttering flight (except in Wood Whites *Leptidea sinapis*), common in most habitats. The 'cabbage whites' are particularly familiar, while Brimstone *Gonepteryx rhamni* and Orange-tip *Anthocharis cardamines* are among our most popular species.

Male Brimstone.

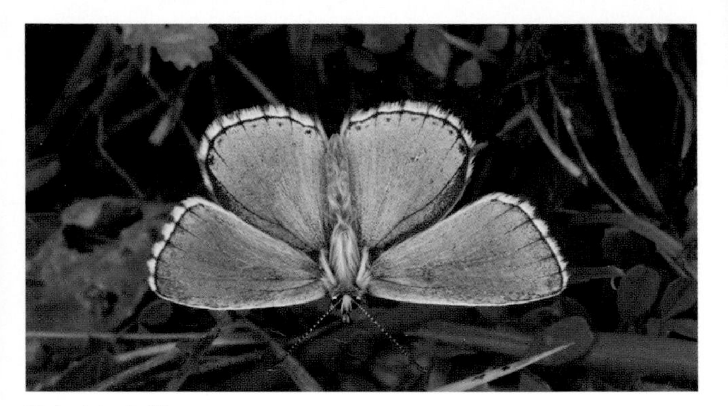

Male Adonis Blue.

Blues, hairstreaks and coppers A large family with 15 or so UK species. They are quite variable in colour but all are small, sun-loving butterflies. Male blue butterflies are mostly blue, the coppers are orange and the hairstreaks have small 'tails' on their hindwings.

Vanessids and admirals Eight medium to large, colourful, strong-flying, mobile species, including a number of far-flying migrants such as Painted Lady *Cynthia cardui*. Usually the upper surface is brightly coloured, while the undersurface is dark, resembling a dead leaf. They are associated with woodland and gardens, and most species visit flowers regularly.

Red Admiral basking.

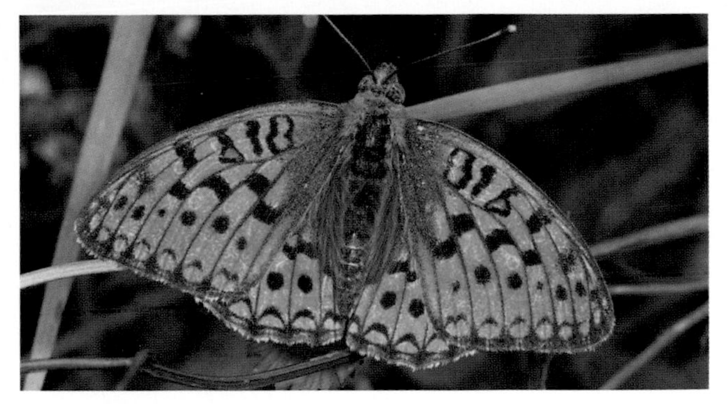

Male Dark Green Fritillary.

Fritillaries Although closely related to the above group, the fritillaries are distinctively orange with black-chequered or latticed markings, and pale-patterned undersides, often with white or silver patches. There are eight species in this group, all loosely associated with woodland and glades.

Browns A rather variable group, formerly in its own family but now included in the huge Nymphalidae together with the above two groups. There are about a dozen species, mostly brownish or orange, and similar above and below, although the group also includes the more conspicuously black and white Marbled White *Melanargia galathea*.

Female Meadow Brown.

Moths

Mint Moth – a micromoth.

Micromoths These include a huge diversity of moths, mostly with forewings less than 10mm long, though the separation from macromoths is artificial. There are over 1,600 species in the UK alone.

Burnets and foresters Small group of day-flying, brightly coloured moths. Burnets are mostly red and black, and have broad, clubbed antennae that are much thicker and more gradually clubbed than those of butterflies. Foresters are shiny bluish-green.

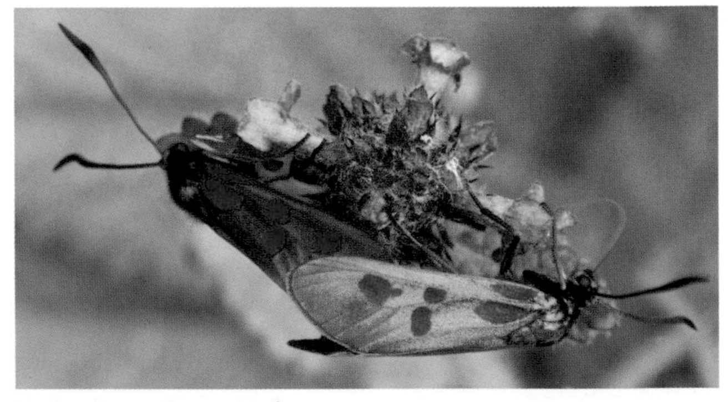

Mating 6-spot Burnet Moths.

Hawkmoths Small group (about 17 species) of large, brightly coloured moths, about half of which are resident in the UK, while the others are regular or occasional visitors. They include the largest British moths. Most have narrow, pointed forewings, which are held slightly swept back while at rest. The larvae of most (but not all) species have a pronounced pointed horn at the rear end.

Narrow-bordered Bee Hawkmoths.

Tiger moths Brightly coloured, bulky, hairy moths, with hairy caterpillars. Most species have very brightly coloured underwings, which are part of their defence against predation.

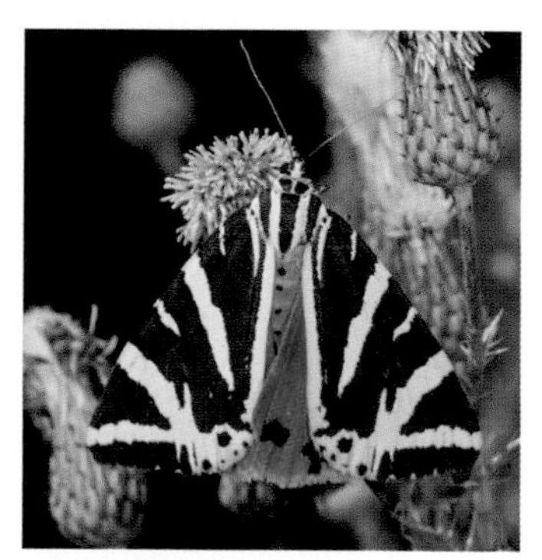

Jersey Tiger Moth.

Geometer moths Distinctive but less familiar large group of species. Adults are often flimsy and weak flying, settling with the wings held completely flat when at rest. The caterpillars are known as 'loopers' due to the way they arch and loop their bodies as they move along. Typical geometers include carpet moths and most of the emeralds.

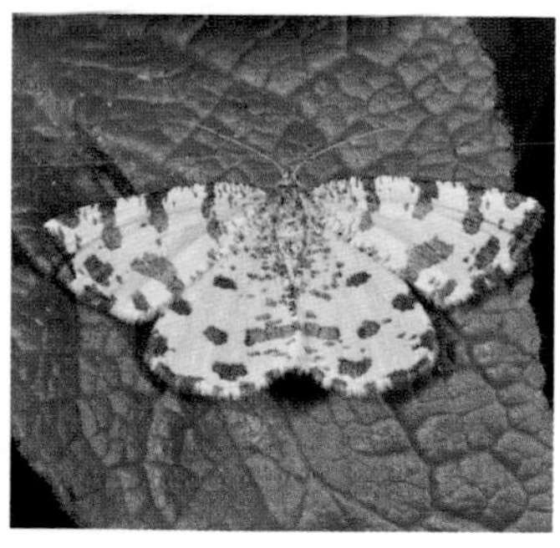

Speckled Yellow Moth.

CADDIS FLIES, TRICHOPTERA

This is a distinctive group of rather moth-like insects, with almost 200 species in Britain. They are best known from their larvae, most of which live in cases made from sand or other debris. They are all aquatic, and the adults are nearly always found near streams, rivers and still water, flying mainly at night. Some species come readily to lights such as moth traps. Caddis flies occur in a wide variety of wet habitats, but particularly along well-oxygenated streams and small rivers, throughout Britain and Ireland, especially in the west.

Trichoptera means 'hairy-winged', and caddis flies are noticeably hairy, especially on the forewings. At rest the long antennae are held pointing forwards, and the wings are held like a pitched roof over the body. In some species the adults may only live for a few days, while in others, such as *Glyphotaelius pellucidus* (illustrated right), they live for several months through the summer, feeding at a variety of flowers.

With a few exceptions, identifying individual species is a specialised job, and most people are happy enough to identify any specimen as a caddis fly.

Above: A caddis fly pupa.
Right: A caddis fly, *Glyphotaelius pellucidus*.

TRUE FLIES, DIPTERA

Although many common names for insects include the word 'fly', such as scorpion fly and caddis fly, the true flies are all in the order Diptera, which is characterised by having only two wings (apart from in a few wingless parasitic species) and sucking mouthparts. The rear pair of wings is replaced by a pair of tiny, pin-shaped bodies known as the halteres, which aid the insect's balance in flight. There are almost 7,000 species of true fly in Britain alone, in a wide range of forms.

The huge eyes of a female horsefly.

Giant Crane-fly

Tipula maxima

The crane-flies, also popularly known as daddy long-legs, are a distinctive smallish group of about 300 species in Britain, easily identified as a group by their large size (although many less familiar ones are quite small), very long, thin, fragile legs, and a V-shaped groove on the thorax. Most individual species are hard to identify, although a few, such as Giant Crane-fly, are distinctive. In this species the dark-spotted wings spread to about 60mm across, while the total leg span can be as much as 100mm.

FLIGHT PERIOD May–August.

HABITAT AND DISTRIBUTION Widespread and quite common throughout the British Isles in damp, shady habitats, especially on acid soil.

SIMILAR SPECIES *T. vittata* is smaller and has wings that are less spotted. *T. oleracea* and *T. paludosa* are the familiar dull brownish crane-flies with unspotted wings, common everywhere.

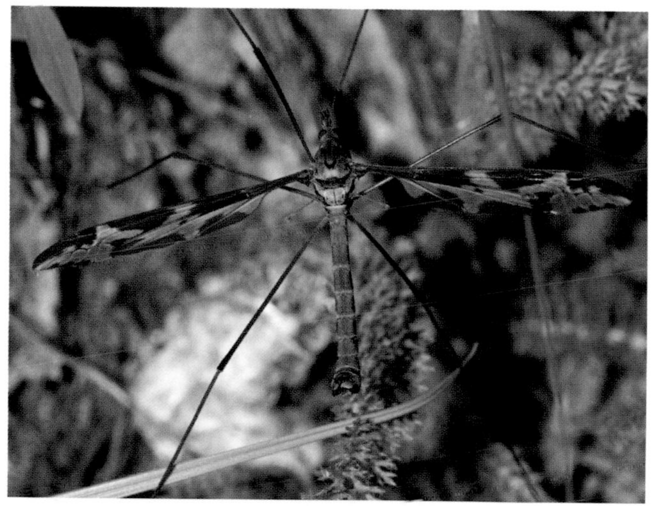

Giant Crane-fly.

St Mark's Fly
Bibio marci

This is a medium-sized, dark black fly up to 15mm long. Both its common name and its scientific name come from its regular appearance around St Mark's Day, 25 April. It has a slow, drifting flight, often at about head height, and frequently settles on both shrubby and herbaceous vegetation. The males have clear wings and large heads, while females have smoky wings and heads that are much narrower than the thorax.

FLIGHT PERIOD April–June, although most common late April–early May.

HABITAT AND DISTRIBUTION Locally common almost throughout the area, except Ireland, along hedges and wood margins, and in rough, scrubby grassland.

SIMILAR SPECIES Two closely related species, *B. hortulanus* and *B. anglicus*, are smaller, and the females have a bright red abdomen. Both are local, and mainly southern.

Female St Mark's Fly.

Snipe-fly
Rhagio scolopacea

This is a medium-sized fly about 10–12mm long, with clear shiny wings that have a few dark markings. Its most distinctive feature is its habit of perching on vertical surfaces such as tree trunks with the head facing downwards, as a result of which it is sometimes known as the 'downlooker fly'. The abdomen is striped orange and black, and tapers markedly towards the tip. Snipe-flies are predatory, making short forays from their vertical perches. The carnivorous larvae live in damp soil.

FLIGHT PERIOD May–August.

HABITAT AND DISTRIBUTION Locally common throughout the area, in damp or shady, well-vegetated places such as woodland margins and glades, and hedgerows.

SIMILAR SPECIES There are several closely related species, differing in body colour and in the amount of wing spotting.

Snipe-fly perched.

Soldier-flies

Stratiomys spp.

This is a small group consisting of about 50 species in the UK. Their name derives from the bright colours of some species, which resemble soldiers' uniforms (slightly). They are slow-flying, broad-bodied flies, most often seen resting on vegetation close to water. The adults visit flowers, and the larvae live in damp soil, compost or water. The species illustrated, *Stratiomys chamaeleon*, sometimes known as the Clubbed General, is probably the most conspicuous UK species, with a broad body and bright yellow and black abdomen, but now also one of the rarest.

FLIGHT PERIOD June–September.

HABITAT AND DISTRIBUTION Local and rare in and around fens and calcareous seepages.

SIMILAR SPECIES There are three other similar species in the genus *Stratiomys*, with slightly different markings; none is common.

A Soldier-fly *Stratiomys chamaeleon*.

Speedwell Gall-midge

Jaapiella veronicae

This insect is tiny, just a few millimetres long, and is normally overlooked even though it is quite common; however, the galls it produces are very conspicuous. The female lays eggs in the buds or shoot-tips of speedwells, especially Germander Speedwell, causing the buds to become large, roughly spherical, very hairy galls. There can be several on one plant, considerably reducing its growth and flowering.

FLIGHT PERIOD Midges may be seen in summer; galls visible from June onwards.

HABITAT AND DISTRIBUTION Widespread and common in grassy, sunny places throughout the area.

SIMILAR SPECIES The galls, if on speedwell, are unmistakeable.

The galls of Speedwell Gall-midge.

Band-eyed Brown Horsefly

Tabanus bromius

This is the most common of several rather similar horseflies that occur in the UK. Horseflies are notable, even notorious, for their strong biting mouthparts, stout bodies and large, brightly coloured eyes. It is the females that have the dagger-like mouthparts, and they feed mainly on blood, while the males feed largely on nectar. They mainly attack domestic and wild large mammals, although many are not averse to biting humans. The larvae are scavengers or predators in marshy or damp places. This species is about 20mm long, although there are both larger and smaller species.

FLIGHT PERIOD June–August.

HABITAT AND DISTRIBUTION Found in and around damp pastures, and confined to the southern parts of Britain and Ireland.

SIMILAR SPECIES Dark Giant Horsefly *T. sudeticus* is even larger. Golden Horsefly *Atylotus fulvus* has golden hairs and green eyes.

Female Band-eyed Brown Horsefly.

Square-spot Deerfly
Chrysops viduatus

The deer-flies are attractive, boldly coloured, triangular-shaped flies about 10–12mm long with conspicuously colourful eyes. There are four rather similar UK species. Their wings are held in a delta shape, and are variably covered with large, smoky patches. Although one's first impressions of them may be favourable, second impressions are unlikely to be – they bite hard and quickly. Square-spot Deerfly has a squarish black spot on the second abdominal segment.

FLIGHT PERIOD June–September.

HABITAT AND DISTRIBUTION Widespread and locally common in rough, damp habitats including heaths, bogs and pastures, mainly in southern Britain.

SIMILAR SPECIES The three other species are distinguishable by small details of eye colour and body markings. *C. relictus* is the most widespread of them.

Female Square-spot Deerfly.

Cleg-fly or Notch-horned Cleg

Haematopota pluvialis

These dull-coloured relatives of the horseflies are much the most likely to bite you; they approach quietly and usually remain unnoticed until they actually bite. They are brownish-grey in colour and about 10mm long, with mottled wings that are held tent-like over the body. As is the case with most other biting flies, it is the females who bite to feed on blood before egg-laying. The larvae are little known and rarely seen, but probably live as scavengers and predators in damp, marshy soil.

FLIGHT PERIOD May–September, peaking June–July.

HABITAT AND DISTRIBUTION Widespread and locally common in rough, damp habitats, including heaths, bogs, wet woodland and pastures, throughout the UK.

SIMILAR SPECIES There are several other similar closely related species, although this is much the most common one.

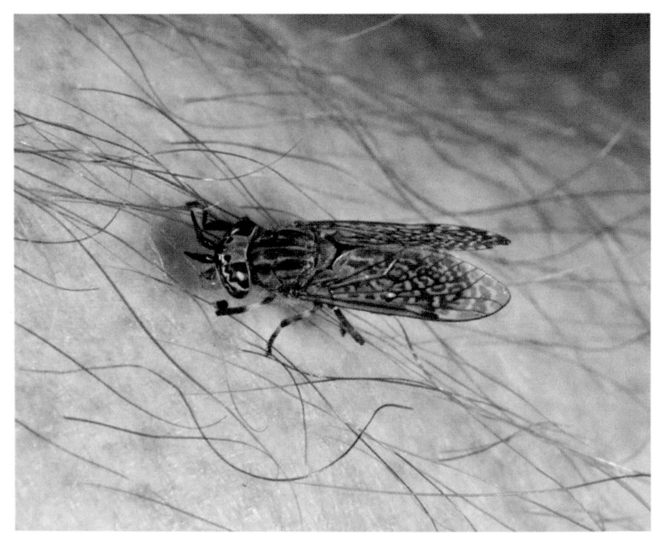

Female Cleg-fly biting.

Bee-fly
Bombylius major

This very distinctive medium-sized fly is about 12mm long, with a brown, roundish, furry body, and a strikingly long proboscis that cannot be coiled or retracted. It spends much time in spring feeding at flowers, especially primroses, by hovering or clinging to the flower. The female scatters eggs on sandy ground, and the resulting larvae seek out, or get carried to, the nests of solitary bees, where they become parasitic on the host's larvae.

FLIGHT PERIOD March–June.

HABITAT AND DISTRIBUTION Widespread throughout the area except in the far north, in warm, flowery places such as gardens and woodland edges.

SIMILAR SPECIES There are eight other species in the UK, but the size and long proboscis of this one is distinctive. Heath Bee-fly *B. minor* is most similar, but much smaller and rarer.

Bee-fly basking.

Hornet Robber-fly
Asilus crabroniformis

This is the largest and most distinctive of the 30 or so species of robber-fly in the UK. These powerful predators are easily recognisable as a group, having cylindrical bodies, piercing beaks and very bristly faces. This species is large, up to 28mm long, and is boldly yellow and black with brownish wings – not unlike a Hornet. It is most commonly found in open areas with dung, where the eggs are laid. The adults use the dung or nearby perches to make forays in search of insects, which they catch and hold with their powerful front legs.

FLIGHT PERIOD June–October.

HABITAT AND DISTRIBUTION Uncommon and probably declining in pastures and heaths with grazing animals, only in Wales and southern England.

SIMILAR SPECIES There are many smaller, less boldly marked robber-flies, but none could be confused with this species.

Mating Hornet Robber-flies.

Common Awl Robber-fly

Neoitamus cyanurus

Although less distinctive than Hornet Robber-fly, this medium-sized robber-fly (up to 20mm long) is reasonably easy to identify. It has a very slender, cylindrical body, a greyish thorax striped with dark grey, a dark brown body expanded at the tip of the abdomen and black and orange legs. Males have a blue patch at the rear and females have an extendible ovipositor.

FLIGHT PERIOD June–September.

HABITAT AND DISTRIBUTION Quite common in England and Wales, but rare elsewhere, occurring particularly in oak woodland.

SIMILAR SPECIES Many other robber-flies look rather similar. Dune Robber-fly *Philonicus albiceps* is probably most similar, but it is a coastal species.

Common Awl Robber-fly.

Stiletto Fly

Spiriverpa lunulata

There are 14 species of stiletto fly in the UK; this is one of the largest and most distinctive. It is an attractive little fly up to 11mm long, with a conspicuous silvery-grey, hairy, tapering body, big red-brown eyes and dark reddish-brown legs. The halteres have dark knobs at the tips, and the species has noticeably long legs. Few details are known of the early stages, although the larvae are known to be predators in the soil.

FLIGHT PERIOD May–August.

HABITAT AND DISTRIBUTION Local in Scotland, Wales and northern England, and particularly associated with river valleys.

SIMILAR SPECIES Coastal Stiletto Fly *Acrosathe (Thereva) annulata* is similar, but has yellow halteres and is mainly coastal.

Male Stiletto Fly.

Glittering Green Fly

Poecilobothrus nobilitatus

This attractive little fly is only 7–8mm long, but is made more conspicuous by its appearance and habits. Both sexes have bright green bodies, smoky wings and two neat rows of hairs on the thorax, but the males also have conspicuous white wing-tips. The flies are most often seen in pairs or groups on pond-side mud or floating vegetation, with the male waving his wings about as if guiding a plane in to land. This courtship display precedes mating.

FLIGHT PERIOD May–September.

HABITAT AND DISTRIBUTION The species is always associated with water, from tiny muddy ponds to slow-flowing rivers, throughout England, Wales and Ireland, and is declining towards the north.

SIMILAR SPECIES There are a number of other 'long-legged flies', but none has the particular combination of characteristics of this one.

Courting Glittering Green Flies, male on left.

Hoverflies

The hoverflies, all in the family Syrphidae, are a large and distinctive group. The 275 species in the UK include many familiar insects, and there are many more elsewhere in Europe. They are characterised by their bright colours, ability to hover regularly and habit of feeding on nectar. Many are wasp mimics brightly banded with black and yellow. The larvae, however, are widely variable in appearance and habits; some are predatory on insects such as aphids, some are aquatic, some are scavengers and others eat solely plant material.

Marmalade Hoverfly

Episyrphus balteatus

This is one of the most common, familiar and distinctive of the hoverflies. The adults have abdomens that are yellow to orange with transverse broad black bands, but behind each of two black bands there is a second thinner band, like an echo. The gap between the bands may often be paler than the rest of the abdomen colour. The larvae of this species are small, slug-like, voracious predators of aphids, and are thus welcome in gardens and on farms.

FLIGHT PERIOD May be seen all year, in any month, but most frequent March–October. There is some evidence that adults may hibernate.

HABITAT AND DISTRIBUTION Widespread and common throughout the British Isles in sunny, flowery habitats such as gardens, meadows and woodland glades. Resident populations may be regularly boosted by huge influxes of immigrants from the south when conditions are right.

SIMILAR SPECIES None. The double black markings of this species are distinctive.

Right: Marmalade Hoverfly on Knapweed.

Pied Hoverfly

Scaeva pyrastri

This large and conspicuous hoverfly is up to about 12mm long, with three pairs of upward-curving white commas on a black abdomen background. Within the UK this is a distinctive combination, unlike that of almost any other species. The adults visit a wide range of flowers, constantly seeking nectar. The larvae are aphid eaters, and although the species breeds in Britain, few individuals survive the winter and most of our population comes from annual immigrations.

FLIGHT PERIOD May–November.

HABITAT AND DISTRIBUTION Widespread and often common throughout the British Isles in sunny, flowery habitats such as gardens, meadows and parks.

SIMILAR SPECIES *S. selenitica* has white marks that are angled downwards and clubbed at the tip; it is uncommon. *Dasysyrphus venustus* is smaller, with yellow comma markings.

Pied Hoverfly on a daisy.

Beaked Hoverfly
Rhingia campestris

This highly distinctive species cannot be confused with anything else (except for one close relative). It is about 10–12mm long with a broad orange body lightly marked with black, and a highly developed, protruding snout that extends well forwards in front of the eyes. This feature allows it to feed in deep-throated flowers such as Bugle. The larvae breed in cow dung and other faeces.

FLIGHT PERIOD Seen at any time in April–November, but most sightings are in May.

HABITAT AND DISTRIBUTION Widespread and common throughout the British Isles, particularly where there are cattle and good nectar sources together.

SIMILAR SPECIES *R. rostrata* is very similar but has no central black line on the abdomen, and no black line along the edges of the abdomen segments. It is much less common.

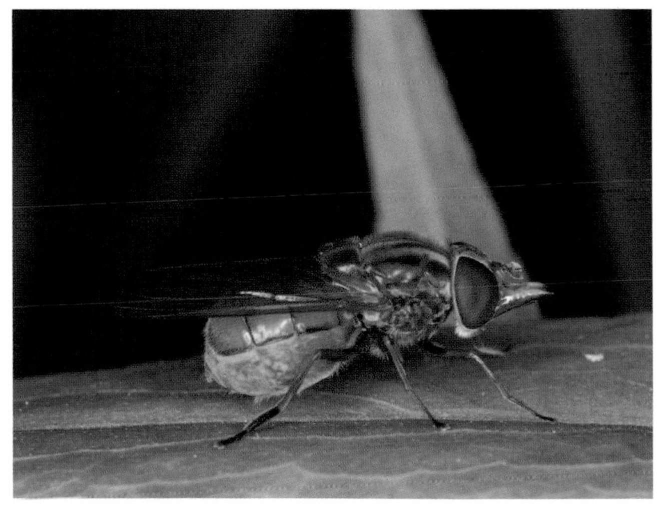

Beaked Hoverfly.

Volucella bombylans and other large hoverflies

This is a large and highly distinctive hoverfly – except that it looks like a bumblebee. It is probably the most accurate of the various bumblebee mimics, and even occurs in different forms mimicking both Red-tailed and White-tailed Bumblebees (see p. 146–7). It has a very furry body, about 16mm long in total, in roughly the right combination of colours for the equivalent bumblebee. It can be distinguished from bumblebees by a combination of its lazier behaviour and frequent basking, the shiny black, hairless thorax, the short antennae, a black spot on the wing and the presence of only one pair of wings. The females lay their eggs in the nests of bumblebees and other hymenoptera, where the larvae develop as scavengers and occasional predators.

FLIGHT PERIOD May–September; most common June–July.

HABITAT AND DISTRIBUTION Widespread and common throughout the British Isles in rough, flowery places such as meadows, woodland margins and scrub.

White-tailed Bumblebee mimic, *Volucella bombylans*.

Volucella pellucens on Dwarf Elder.

SIMILAR SPECIES Other species of *Volucella* are at least as large, but none are so hairy. *V. pellucens* is hairless with a boldly black and white abdomen, and is common around woods. *V. inanis* is slightly Hornet-like and striped orange and black, and is mainly southern around woods and gardens. Another bee mimic, *Pocota personata*, differs in having the front half of the thorax covered in orange-brown hairs, and the rear part shiny black. It is uncommon and mainly southern.

Volucella inanis on Blackberry flower.

Drone-flies
Eristalis spp.

The drone-flies comprise a very abundant group of hoverflies consisting of about 10 species in the UK. This species somewhat resembles a Honey Bee (hence the name – a drone is a male Honey Bee) in general size and colour, and lazily visits flowers for nectar, especially in late summer and autumn. It differs from Honey Bee in having only two wings and much shorter antennae, and in details of its markings. The remaining common species in this group can be quite hard to distinguish from each other – requiring detailed examination of face and leg colours – but are quite distinctive as a group. They have quite a dark, furry thorax and abdomen, with broad, variable dull yellow markings coming in from each side (but not meeting in the middle), and two or three thin white bands across the abdomen.

The courtship of this group is fascinating, as one or more males hover above a female, which is usually on a flower,

Eristalis interruptus.

Drone-fly *Eristalis tenax.*

producing a specific wing noise that is presumed to be part of the ceremony. The female gradually becomes interested, although curiously mating is very rarely observed. The larvae of this group are mainly aquatic or found in wet mud, and some are known as rat-tailed maggots because of their shape.

FLIGHT PERIOD Most frequent April–September, but drone-flies have been seen in every month.

HABITAT AND DISTRIBUTION Widespread and common throughout the British Isles in rough, flowery places such as meadows, woodland margins and scrub.

SIMILAR SPECIES *Helophilus pendulus* is similar, but has a conspicuous longitudinally striped, black and yellow thorax – it is sometimes called 'the footballer' for this reason. It is common everywhere there are flowers.

Helophilus pendulus.

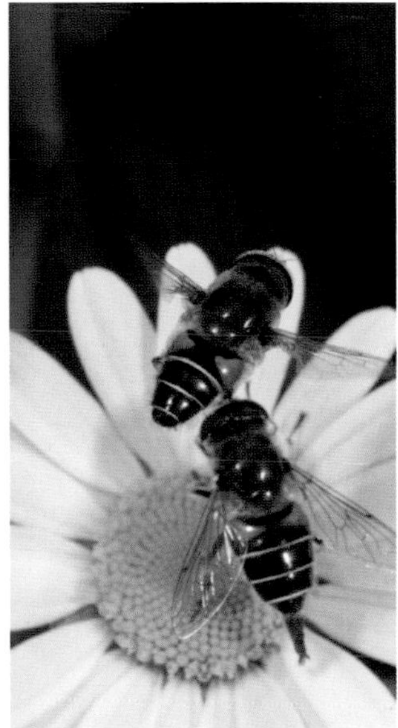

Courting *Eristalis interruptus.*

Picture-winged flies

This is a small group of flies in the family Tephritidae. They are quite small (6–8mm long), but have attractively 'painted' wings and black or orange bodies. Their larvae develop in the buds or stems of plants, mainly thistles, where they cause galls to form as they develop by the secretion of growth hormones. The most visible of these is caused by *Urophora cardui*, which produces quite large, hard, swollen green galls up to 60mm long, on the stems of thistles, especially Creeping Thistle. The adult flies are 6–8mm long and dark-bodied, with sinuously dark-marked wings.

By contrast, the most commonly seen adult fly is *Xyphosia miliaria*, which is observed most frequently on the flowers and buds of Marsh Thistle. The flies have bright orange-red heads and bodies, are up to 7mm long and have clear wings with black marks. Females have a dark, tapered tip to the abdomen. The galls they cause are smaller and rarely noticed.

FLIGHT PERIOD Most frequent as adults May–August, depending on the species.

HABITAT AND DISTRIBUTION
Widespread and locally common in southern UK, in rough, flowery places with thistles.

SIMILAR SPECIES There are many other similar flies, but these two are the most frequent and distinctive.

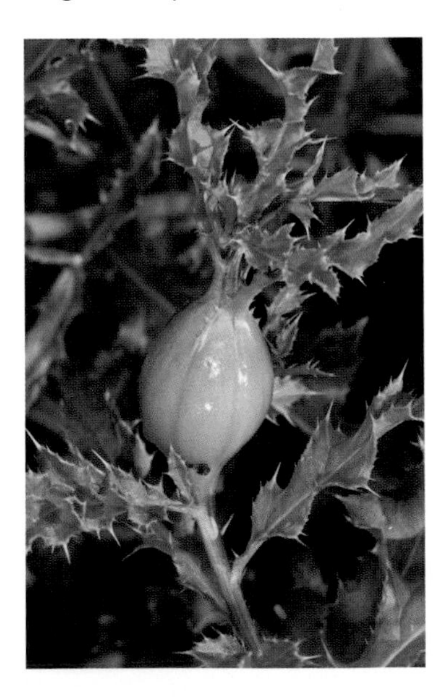

Right: Gall of *Urophora cardui*.

Opposite: *Xyphosia milaria* on Marsh Thistle.

Parasitic tachinid fly

Tachina fera

This medium-large fly (up to 15mm long) is one of more than 260 species of tachinid fly (Tachinidae family) in the UK. It has a broad orange body with a central black stripe on the thorax, all covered with thick black bristles. The legs are mainly reddish-brown, and the bases of the wings are yellowish. The females lay eggs in or close to caterpillars of moths and other insects, which the larvae then eat – this life-style is known as parasitoid.

FLIGHT PERIOD April–September.

HABITAT AND DISTRIBUTION Widespread and locally common throughout the UK in damp, flowery, unspoiled habitats.

SIMILAR SPECIES *T. grossa* is larger (up to 20mm long), and all black except for a yellow face, with blacker legs. It is less common, and occurs mainly in the south.

Tachina fera on Angelica.

Flesh Fly
Sarcophaga carnaria

This medium-large fly of the Sarcophagidae family is up to 15mm long, and has red eyes, large feet and a black and white chequered abdomen. It breeds in carrion of various sorts, in which the females, unusually, lay live larvae rather than eggs. This speeds up the life-cycle in what is a transient and changing environment, allowing the next generation to emerge before the carrion disappears.

FLIGHT PERIOD May be seen throughout the year except in very cold weather.

HABITAT AND DISTRIBUTION Widespread and common throughout the UK in most habitats, and often common around houses.

SIMILAR SPECIES There are several closely related species, which are hard to separate without microscopical examination.

Flesh Fly.

Urban Bluebottle

Calliphora vicina

This is probably the most common of a small group of similar species. They are medium-sized, bristly flies, up to 12mm long, with broad, compact, dark bluish-black bodies, and a slight metallic tinge to the abdomen. This species has reddish jowls and a small yellow patch at the base of the wing. The eggs are laid in carrion or meat-origin food, frequently in houses. Their presence in corpses has proved very useful in forensic examinations, helping to date the time of death. They can often be seen resting on sunny walls and posts.

FLIGHT PERIOD Seen throughout the year except in very cold weather.

HABITAT AND DISTRIBUTION Widespread and common throughout the UK in most habitats, especially around houses.

SIMILAR SPECIES *C. vomitoria* is similar, but has yellowish hairs on its face and no yellow on the wings. Greenbottles *Lucilia caesar* are bright metallic green all over.

Urban Bluebottle.

Cluster-fly
Pollenia rudis

This smallish fly (up to 10mm long) would probably normally be overlooked, aside from its habit of gathering in masses in autumn, then entering houses in swarms to hibernate. The flies are often only noticed when they try to emerge on the first warm days of spring. They also hibernate in old nests, tunnels and hollow trees. On close examination they are very distinctive, with long golden hairs on the abdomen, and a black and white chequered thorax. The larvae are parasites of earthworms.

FLIGHT PERIOD Seen throughout much of the year except in very cold weather, but most noticeable in autumn.

HABITAT AND DISTRIBUTION Widespread and common throughout the UK in most habitats, especially around houses.

SIMILAR SPECIES There are three less common species in the genus.

Cluster-fly in autumn.

Noon Fly

Mesembrina meridiana

This common and distinctive, medium-sized fly is up to 12mm long, and readily recognised by its black, hairy body and conspicuous yellowish-brown wing-bases. On warm days in autumn the flies gather – often in quite large numbers – to sunbathe on fence posts, tree trunks and other warm surfaces. The eggs are laid in dung, but the larvae are not dung eating; they are carnivorous, feeding on the larvae of other insects. Females lay only one egg per dung mass.

FLIGHT PERIOD April–October.

HABITAT AND DISTRIBUTION Widespread and common throughout the UK in most habitats, especially close to grazing pastures.

SIMILAR SPECIES Some tachinid flies (see p. 114) have a similar wing pattern, but they are much more bristly.

Clustered Noon flies.

Yellow Dung Fly
Scathophaga stercoraria

This is a common and familiar fly in rural areas, gathering in large quantities wherever there is cow dung, and less commonly on other types of dung. The males are slender-bodied, up to 10mm long and covered in golden-yellow hairs. The females are less hairy, greyer or greenish-brown, and much less frequently seen because they only visit dung to mate and lay eggs. The males gather on dung, both to attract females and to prey on other dung-visiting insects such as blow-flies. Both sexes also feed on

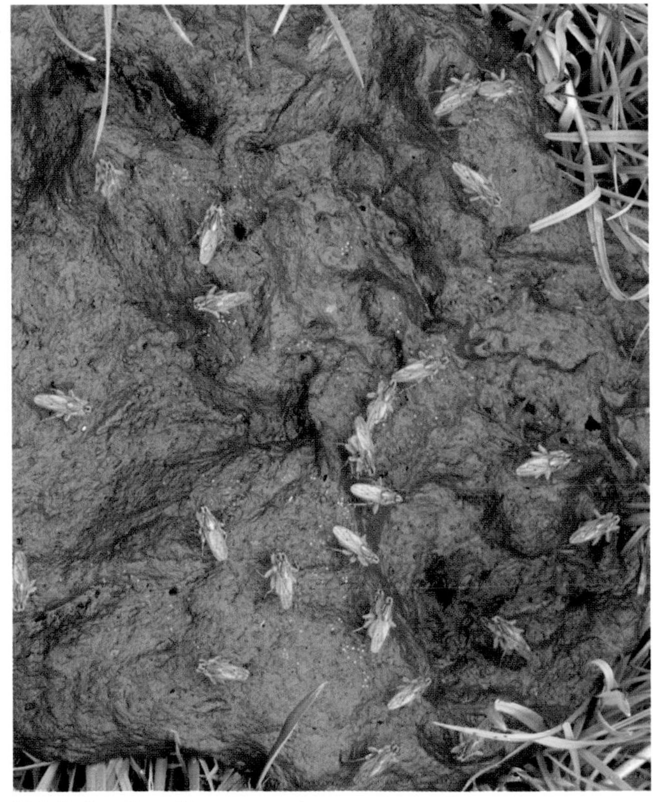

Male Yellow Dung flies on cow dung.

Male Yellow Dung-fly.

nectar and dung. The eggs are laid in the dung and the larvae feed on it. In some areas this species has declined, due partly to the use of pesticides on cattle farms, although it is still widespread and common throughout most of the northern hemisphere. It is a commonly used laboratory subject because of its short life-cycle.

FLIGHT PERIOD April–September.

HABITAT AND DISTRIBUTION Widespread and common throughout the UK in and around cattle- and horse-grazing pastures.

SIMILAR SPECIES *S. furcata* is similar but more slender, and has dark patches on the wings.

Thick-headed Fly
Myopa buccata

The thick-headed, or conopid, flies are a specialised group of about 25 species in the UK. This species has a hunched brownish body, slightly smoky wings and short antennae (compared with *Conops* species, which have quite long antennae). Thick-headed flies have a specialised life-cycle: fertile females aggressively intercept their host species (usually a bumblebee for this species, but others attack solitary bees or wasps) and using the modified abdomen, they slit open the host's body to lay an egg into it. The larva then develops in the host's body. Adults feed on nectar at various flowers.

FLIGHT PERIOD April–August.

HABITAT AND DISTRIBUTION Widespread throughout the UK in rough, flowery habitats, wherever the host insects occur.

SIMILAR SPECIES There are several other *Myopa* species. The most frequent is *M. testacea*, which has a much blacker thorax than *M. buccata*.

Thick-headed Flies mating.

SAWFLIES, ANTS, WASPS, BEES AND RELATIVES, HYMENOPTERA

This is an enormous order of insects with about 7,500 species in Britain and vastly more throughout the world. They vary from tiny insects to huge ones, with a great range of life-cycles and morphology. As a result it is hard to characterise them easily. Essentially, more or less all of them have two pairs of clear membranous wings that are held together by tiny hooks, and biting mouthparts, although these may be much modified.

There are two major suborders within the Hymenoptera – the Symphyta, which includes sawflies, and the Apocrita, which includes bees, wasps, ants and all the remainder. The Apocrita all have the more familiar hymenopteran feature of a marked 'waist' between the thorax and abdomen, whereas the Symphyta do not. Sawfly larvae are mobile with legs, very like moth caterpillars, whereas all the others have legless larvae that live surrounded by their food.

Large Rose Sawfly larvae.

Sawflies and Wood wasps

There are almost 500 species in this suborder (Symphyta), with many varied forms. All have the features described above, and many have a saw-like ovipositor (hence their name).

Honeysuckle Sawfly

Zaraea fasciata

A bulky, largish insect up to 12–14mm long, the Honeysuckle Sawfly has long, clubbed antennae. Both sexes are blackish tinted with violet and have dark patches on the wings, but the females have a clear band of white at the back of the thorax. In fact, males are rarely seen, and most reproduction seems to be parthenogenetic – that is, without fertilisation of the eggs by a male. The eggs are laid in honeysuckle and snowberry, which the larvae eat.

FLIGHT PERIOD April–August.

HABITAT AND DISTRIBUTION Widespread but never common, in wooded areas, parks and gardens, where the food plants occur.

SIMILAR SPECIES Hawthorn Sawfly *Trichiosoma lucorum* is larger, to 20mm in length, browner and has smaller wing markings.

Female Honeysuckle Sawfly.

Gall-forming wasps

This is a family of tiny wasps (Cynipidae), of which about 40 cause galls. The adults are very small and hard to find, let alone identify, but the galls they cause are large and distinctive. The female lays eggs in the tissue of the host plant, often a quite specific species, and as the larvae develop they influence the development of the plant to form a protective gall around themselves. The plant tissue also provides food for the developing larvae. Many species in this group have complex life-cycles, and some have alternating generations that produce quite different galls at different stages, often on different plants.

Robin's Pincushion, or Bedeguar Gall, is one of the most familiar. It is caused by *Diplolepis rosae* on roses, where the larvae cause an outgrowth of masses of reddish branched shoots in a roughly spherical shape. It is common almost throughout the area during the summer.

Robin's Pincushion on rose.

Cherry Galls on oak leaf.

Cherry Galls are hard, spherical, reddish, cherry-like structures found on the undersides of oak leaves in autumn as the leaves fall, and are caused by *Cynips quercusfolii*. It is common throughout the area, and mainly seen in autumn, although there is a less conspicuous spring generation.

Oak Apples, caused by *Biorrhiza pallida*, are among the largest galls in the UK. They are up to 50mm across and are formed on oak branches in spring. The spongy, irregularly shaped 'apple' produces both male and female gall wasps, and starts to collapse by midsummer.

Knopper Galls are large, irregularly fluted, hard outgrowths from acorns, caused by *Andricus quercuscalicis*. This species spread into Britain relatively recently and is now common.

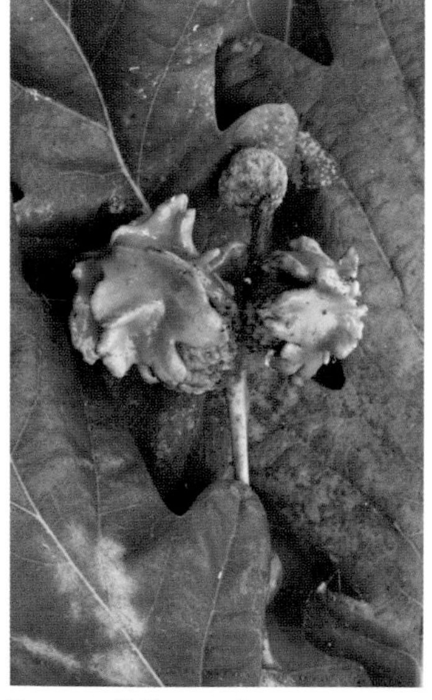

Knopper Galls on oak.

Ichneumons

The ichneumons are an enormous family (Ichneumonidae) of parasitic hymenopterans, with over 3,000 species in the UK if the closely related braconids are included. They vary widely in size from small and sometimes wingless to quite large, with long antennae, and most females have a long ovipositor. Most of them parasitise the caterpillars of moths and butterflies, laying one or more eggs in the hapless host, which then gradually dies as it is eaten from within.

Female Commander Ichneumon egg-laying.

Sabre Wasp or Giant Ichneumon

Rhyssa persuasoria

Sabre Wasp is an extraordinary insect with a total length (including the ovipositor) of 80–100mm. The legs are red, and the body is long and slender. It is black with white marks and terminates (in the female) in a very long, hair-like ovipositor. Males are rather smaller. The female uses her antennae to detect the larvae of horntails (a type of sawfly) or longhorn beetles buried deep within a tree trunk. She then drills down precisely through the wood to find the larva and lay an egg on it. The developing wasp larva then feeds on its host externally.

FLIGHT PERIOD July–August.

HABITAT AND DISTRIBUTION Widespread but very local, mainly in pine woods and plantations.

SIMILAR SPECIES *Ephialtes manifestator* is rather smaller and does not have the boldly marked abdomen. It is widespread but uncommon, and a parasite on longhorn beetle larvae.

Female Sabre Wasp egg-laying.

Yellow Ophion

Ophion luteus

This medium-large ichneumon, up to 20mm long, has a narrowly cylindrical, orange-red body with similarly coloured legs and antennae, clear, black-veined wings and dark blackish eyes. It is mainly active at night and often comes to lit windows or moth traps. The females lay their eggs in caterpillars using the long, thin ovipositor, which may also be utilised as a defence mechanism.

FLIGHT PERIOD Seen at any time in March–September, but most often in late summer.

HABITAT AND DISTRIBUTION Widespread and common throughout the UK, in woods and gardens.

SIMILAR SPECIES *Netelia testacea* is similar in size and colour, but has a black tip to the abdomen.

Yellow Ophion ichneumon.

Ruby-tailed wasp

Chrysis ignita

This small but very brightly coloured insect, up to 14mm long, is also known as the Jewel Wasp. The head and thorax are bright metallic greenish-blue, and the abdomen is metallic red-purple – a striking combination. This and several closely related species are cuckoo wasps (family Chrysididae) – that is, they seek out the nests of mason bees and other solitary bees, in which they lay their eggs. They are most often observed on banks and walls, investigating nest-holes.

FLIGHT PERIOD May–September.

HABITAT AND DISTRIBUTION Widespread and common throughout the UK in most habitats where solitary bees nest, including heaths, gardens and downland.

SIMILAR SPECIES There are several similar species, which may need detailed examination for identification. *C. fulgida* is relatively easy as it has a distinct blue-green section at the front of the abdomen.

Female Ruby-tailed Wasp seeking nest-holes.

Ants

There are about 50 species of these familiar insects in the UK. As a family (Formicidae) they may be recognised by their elbowed antennae, very narrow waist and long thorax, and most of those we see are flightless. They are social insects living in colonies with one or more queens. Most of the ants are wingless females called workers, but at the appropriate time new queens and winged males are produced ('flying ants'), which take to the air and mate. The males die soon after and the mated queens return to earth, break off their wings and become egg-laying queens.

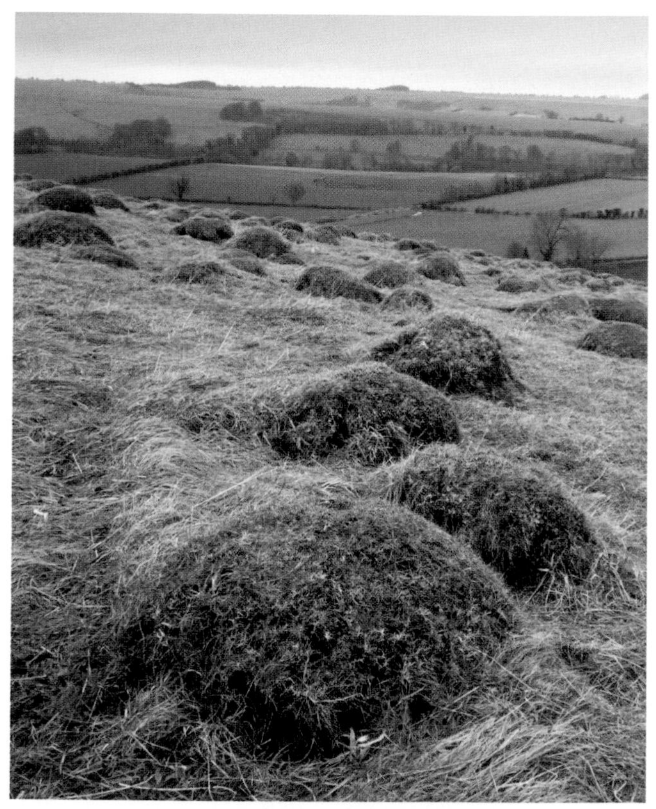

Yellow Meadow Ant anthills in old grassland.

Yellow Meadow Ant

Lasius flavus

This ant is perhaps best known for the products of its activities, rather than for its appearance. The colonies live in open grazed grassland, where they produce perennial anthills that steadily increase in size over the years and may eventually reach 50cm or more in height (ancient grassland can be aged by the size of the largest anthills on it). The ants themselves are about 4mm long, and yellowish-brown in colour all over. They produce enormous mating swarms on warm days in high summer.

ACTIVITY PERIOD Ants and anthills are visible all year round, but flying ants are most likely to be seen July–August.

HABITAT AND DISTRIBUTION Widespread and common throughout the UK, in dry, open, unshaded and unploughed grassland such as chalk downland, where the anthills may become very abundant.

SIMILAR SPECIES Red ants such as *Myrmica rubra* are similar in appearance, but they do not form anthills.

Worker Yellow Meadow Ants.

Southern Wood-ant

Formica rufa

This is another ant species that is as well known for its constructions as its appearance. These wood ants build enormous nests of pine needles and other material (they can be up to 1m high), which are carefully constructed to keep the interiors warm and dry. A large colony may house 250,000 or more ants, and colonies may be spread over several nests. The ant workers themselves are large, about 10mm long, with a black head and abdomen but a red thorax. The workers fan out over their territory in long columns, collecting and killing vast quantities of invertebrates and plant material.

ACTIVITY PERIOD Most likely to be seen from March onwards; mating flights occur in midsummer.

HABITAT AND DISTRIBUTION
Widespread and locally common in sunny woodland, especially coniferous woodland, throughout England and Wales.

SIMILAR SPECIES There are two other species of wood ant in Britain, *F. lugubris*, which is hairier, and *F. aquilonia*, which is more or less confined to Scotland.

Southern Wood-ant workers.

Southern Wood-ant nest.

Wasps

Bee-killer Wasp

Philanthus triangulum

The Bee-killer or Bee-wolf is a striking insect measuring up to 18mm long, with a dark head and thorax, and a bright yellow abdomen variably barred with black bands. It is a voracious predator of Honey Bees (and occasionally other bees), which it catches, paralyses and brings back to its nest burrow in a sandy bank. In some parts of the world it is known to have a significant effect on Honey Bee populations. An egg is laid on each bee, which becomes food for the developing larvae. The adults also visit flowers for nectar. The nest-holes have a distinct triangular shape.

FLIGHT PERIOD July–September.

HABITAT AND DISTRIBUTION Most often found on heaths and sand-dunes, although new habitats are being colonised. Twenty years ago this species was rare and local, but it has greatly extended its range throughout England and eastern Wales.

SIMILAR SPECIES None.

Female Bee-killer Wasp.

Sand-wasp

Ammophila sabulosa

The sand-wasps are highly distinctive insects. They have long, thin bodies up to 25mm long, with a black head and thorax, and a very long, thin, two-segment waist that is red and gradually expands to the slightly bulbous, bluish-black tail. They excavate long, thin burrows in sandy soil, then go in search of caterpillars; these are paralysed, then dragged or carried back to the nest-hole and pushed inside to have an egg laid on them. This acts as food for the developing larva.

FLIGHT PERIOD June–September.

HABITAT AND DISTRIBUTION Occurs on heaths, sand-dunes and other sandy places. Locally common in England and Wales, but absent elsewhere.

SIMILAR SPECIES *A. pubescens* is smaller with a jet-black 'tail', and is confined to south-east England. *Podalonia* species are similar but have a short waist that widens abruptly. They are mainly coastal.

Sand-wasp carrying caterpillar to nest-hole.

Hornet
Vespa crabro

Although the Hornet is essentially a social wasp, it looks very different from wasps and is in a different genus (see also below). It is much the largest of the wasps, with queens measuring up to 35mm long. The head, thorax and first part of the abdomen are shiny reddish-brown, and the rest of the abdomen is like that of other wasps. Hornets are very active predators, visiting flowers mainly in order to catch other insects, which they do very skilfully. Despite their size and powerful sting, they are normally not aggressive unless the nest is threatened.

FLIGHT PERIOD May–October.

HABITAT AND DISTRIBUTION Widespread but local in southern Britain only, and most frequent in old woodland, where they use hollow trees.

SIMILAR SPECIES Other wasps, but all are much smaller and more yellowish.

Hornet, in autumn.

Social wasps
Vespula spp.

The social wasps are a common and familiar, but not entirely popular, group of insects. They are characterised by a boldly black and yellow striped abdomen, a very marked 'waist' between the thorax and abdomen, the way in which the wings are folded back along the sides of the body, and their strong flight and movements. There are nine species in Britain, with broadly similar habits, which are distinguishable by close examination.

Among these social insects each colony starts with an overwintered queen in spring, and she gradually builds the nest and feeds the developing larvae on masticated insects. The pace of the colony expansion increases as workers (sterile females) are produced, which can take over feeding of the new larvae. Eventually a large nest may support up to 10,000 insects. Males are produced later in the year. Towards late summer the insects switch more to feeding themselves on fruit and nectar, which is when they can come into contact with humans. Eventually all but the young queens die, and these then overwinter.

The two most common species are **Common Wasp** *Vespula vulgaris* and **German Wasp** *V. germanica*. They are very similar in appearance and habits, and distinguishable by different markings

Common Wasp.

on the face and body. Common Wasp has a large, anchor-like black mark on the face, whereas German Wasp has a triangle of three dots, although both may vary. The nests of Common Wasp tend to be yellowish because they are made from fresh wood, whereas those of German Wasp are greyer due to them being constructed from older wood. The remaining species are less frequently encountered.

FLIGHT PERIOD Queens first appear in early spring, and individuals may be seen from then until the first hard frosts.

HABITAT AND DISTRIBUTION The two common species may be found almost anywhere.

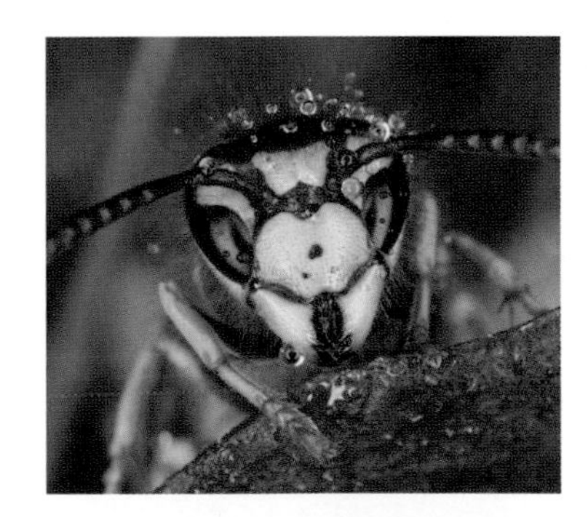

Right: German Wasp face with three dots.

Common Wasp face with anchor mask.

Ivy Bee
Colletes hederae

Ivy Bee is a large and attractive solitary bee up to 13mm long, with a tawny, hairy thorax and a yellow and black abdomen, both of which fade as they age. They are very late emergers, and feed especially on the pollen of ivy in autumn. The males are very active and visible, constantly mobbing females.

FLIGHT PERIOD September–November.

HABITAT AND DISTRIBUTION Most frequent around sandy cliffs and banks, especially on the coast, where there may be many thousands at a site. First recorded in Britain in 2001, Ivy Bees are now locally abundant in the south, especially near the coast, and still spreading.

SIMILAR SPECIES Other *Colletes* species are similar, but use different flowers and are often earlier.

Ivy Bees around cliff nest-hole.

Long-horned Bee

Eucera longicornis

A very attractive and distinctive solitary insect, Long-horned Bee has an extremely furry gingery or greyish body, and very long antennae. The males have the longest antennae, but even those of the female are longer than those of other bees. The bees spend much of their time visiting a wide range of flowers to collect pollen. They are specialist pollinators of some of the *Ophrys* group of orchids, particularly in southern mainland Europe, taking part in the process of pseudocopulation in which the male bee mistakes the flower for a female.

FLIGHT PERIOD May–late July.

HABITAT AND DISTRIBUTION Local in southern Britain only, and most frequent in coastal grasslands, or inland on heaths and in woodland.

SIMILAR SPECIES Males are unlikely to be confused with anything else.

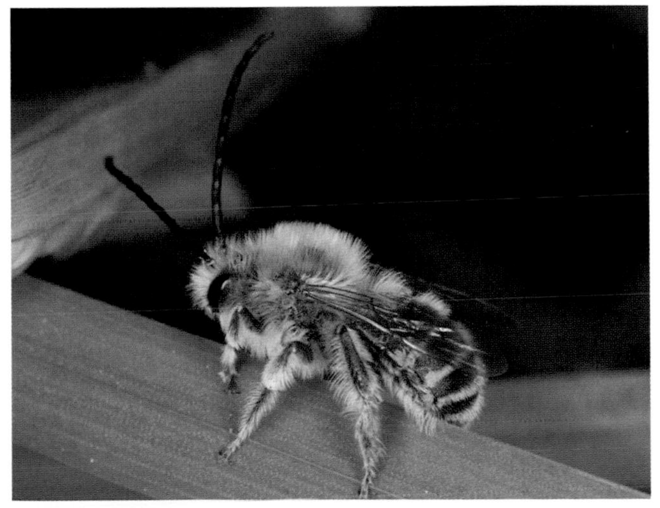

Long-horned Bee.

Wool Carder Bee

Anthidium manicatum

The numerous solitary bee species are notoriously difficult to identify, although this one is relatively distinctive. It is a largish dark species up to 15mm long, with conspicuous yellow spots on each segment down the abdomen. Males are larger than females, and they take up a territory on a suitable plant, such as Downy Woundwort, from which all except females are aggressively driven off. Females cut and comb hairs from suitable plants (hence the name), and use them to line the nests.

FLIGHT PERIOD May–August.

HABITAT AND DISTRIBUTION Quite common in southern Britain only, in gardens, woodland rides, meadows and other flowery, open habitats.

SIMILAR SPECIES Other similar species, such as *A. variegatum*, occur in Europe, but none is found in Britain.

Wool Carder Bee settled.

Cuckoo Bee

Nomada lathburiana

The cuckoo bees are so-called because they behave rather like cuckoos – that is, they lay eggs in the nests of host mining bees. The egg hatches into a larva with very large jaws, which kills the host egg or larva, then waits for the host mining bee to feed it. In fact, cuckoo bees resemble wasps more than bees, having a distinct waist and a brightly banded yellow and black abdomen, with red on the thorax. This species is about 12mm long, and is specific to the mining bee *Andrena cineraria*.

FLIGHT PERIOD April–June.

HABITAT AND DISTRIBUTION Local in southern Britain only, wherever the host occurs, usually in warm, sandy places. Formerly rare, it now seems to be spreading.

SIMILAR SPECIES There are about 30 species in this genus, requiring detailed examination or knowledge of their host species for identification.

A Cuckoo Bee, *Nomada lathburiana*.

Violet Carpenter Bee

Xylocopa violacea

Violet Carpenter Bee is one of the largest bees in Europe, reaching a length of up to 22mm, with a broad, robust body. The body is dark blue-black or slightly brownish; the smoky wings appear either violet or brown according to the light. The males have two red segments near the tips of the antennae. The species nests in old timbers.

FLIGHT PERIOD Overwintered individuals fly April–May; the next generation flies from July until hibernation.

HABITAT AND DISTRIBUTION At present local in southern Britain only, in various flowery habitats. It has been steadily spreading northwards, and now appears regularly in Britain and seems to have begun to breed and overwinter here. It is common in adjacent continental Europe, southwards.

SIMILAR SPECIES **X. valga* is very similar, but the males have all-black antennae; not yet regular in Britain, but occurs in nearby mainland Europe.

Violet Carpenter Bee on vetch.

Leaf-cutter Bee

Megachile centuncularis

The adults of this species are rather undistinguished, looking like many other bees. It is their nesting habits and the results of these that make them so distinctive. The bees are furry, brownish-black and about 12–15mm long, with a yellow pollen brush under the abdomen of the female. Females cut roughly semi-circular discs of about 15mm diameter from leaves, especially those of roses and tutsan. They roll these and carry them back to the nest tunnel, where they are glued together with saliva to form nest cells. The holes in the leaves are distinctive and an immediate indication of the presence of the bees. Adults feed mainly on nectar.

FLIGHT PERIOD April–August.

HABITAT AND DISTRIBUTION Common throughout the UK in gardens, woods and parks.

SIMILAR SPECIES There are several other species in the genus.

Female Leaf-cutter Bee carrying leaf disc.

Honey Bee

Apis mellifera

Honey Bees are among the most familiar of our insects (although they may also be easily confused with hoverflies and other insects). Workers are about 12mm long, with queens and drones being rather larger, up to 16mm long. They are a warm brownish colour with black rings on the abdomen, although they are highly variable due to the large amount of imported genetic material in hive bees. They live in large colonies both in hives and in the wild, with up to 80,000 bees inhabiting a large and healthy colony. The male bees, or drones, appear towards the end of summer – they tend to be fatter and have longer antennae than the workers, and have no function other than to mate. The female workers do all the pollen and nectar collection, and the housekeeping of the hive. Queens survive the winter and may live for several years. Bees sting quite readily and unlike wasps they cannot remove the sting from mammalian skin, so tend to die afterwards.

FLIGHT PERIOD March–October; longer in favourable conditions.

HABITAT AND DISTRIBUTION Widespread and common anywhere there are flowers.

A rural bee-keeper's yard.

Honey Bees drinking.

SIMILAR SPECIES Drone-fly *Eristalis tenax* can look similar, but has only two wings and shorter antennae, and displays sluggish behaviour.

Honey Bee visiting Corn Cockle.

Bumblebees
Bombus spp.

Bumblebees are a distinctive and familiar family (Apidae) of insects made up, in the UK, of 16 or more species, together with a number of closely related cuckoo bees (see p. 148). They are large, robust, furry insects, and are often predominantly black with bands of yellow, white or red. Most workers are about 15mm long, with queens being rather larger. They are social insects, but live in much smaller colonies than Honey Bees. An overwintered queen will start a colony in spring, which will build up to containing 50–150 insects by late summer. The colonies are annual, and only the mated new queens survive through winter.

Identification of bumblebees, with a few exceptions, is notoriously difficult in the field, due to the variation between queens, workers and males, variation within species and the gradual loss of key colours through the season.

FLIGHT PERIOD Most of the bumblebees can be seen in March–October, but numbers only build up from late June onwards.

Red-tailed Bumblebee *Bombus lapidarius* is one of the most distinctive species, at least when it comes to the queens and workers, as they are largely black with a red tail to the abdomen. It is common almost throughout the UK.

Male Red-tailed Bumblebee on Corn Marigold.

Buff-tailed Bumblebee on pansy.

White-tailed Bumblebee *B. lucorum* is usually distinctive, with a lemon-yellow collar and second abdominal segment, and a white tail. Males may be differently patterned, but always have yellow hairs on the face.

Common Carder-bee *B. pascuorum* is rather small and almost entirely covered with ginger-brown hairs except for a few black patches on the abdomen. It is common in gardens and many other habitats.

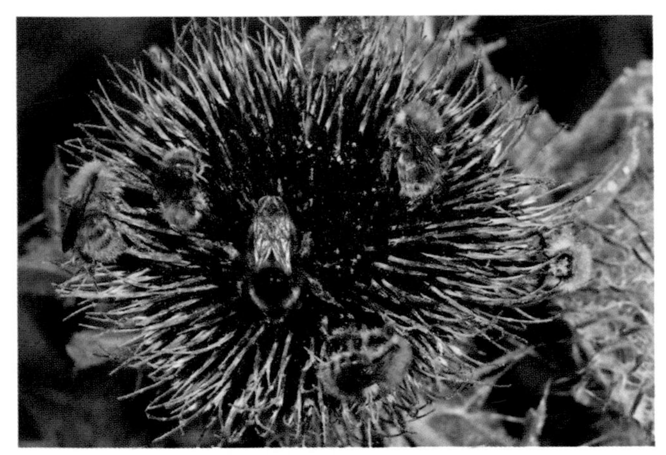

Bees, mainly Carder-bees, on Woolly Thistle.

Cuckoo bees

Bombus spp.

The cuckoo bees were formerly placed in a separate genus, *Psithyrus*, but they are now considered to be just bumblebees, *Bombus* species, that have gone bad! Their life-cycle differs in that they do not build their own nests. The powerful female cuckoo bee emerges later from hibernation, when bumblebees have already started their nests; she then invades the nest of the appropriate species, kills or evicts the queen and lays her own eggs to be looked after by the workers. Thus cuckoo bees produce no workers, and have no pollen baskets.

In appearance cuckoo bees mimic their hosts, but differ in having less hairy bodies that are shinier as a consequence – the black shiny chitin of the thorax often shows through – with smokier, darker wings, and a generally slower flight and activity pattern.

There are six UK species of cuckoo bee. Typical species include *Bombus rupestris*, which is very like its host, Red-tailed Bumblebee, although it is much less common; *B. barbutellus*, which uses and resembles *B. hortorum*, differing in the yellow hairs on top of its head, in addition to the group features mentioned above; and *B. vestalis*, which resembles *B. terrestris* although it lacks a yellow band at the front of the abdomen.

Cuckoo bees are commonly on the wing in May–October, in the same habitats as bumblebees.

A Cuckoo Bee, *Bombus vestalis*.

A Red-tailed Cuckoo Bee, *Bombus rupestris*.

BEETLES, COLEOPTERA

The beetle order is the largest in the world, with something like 375,000 species worldwide. In Britain there are a mere 4,000 or so species, which is enough to encompass a wide variety of sizes and forms, from minute up to the giant Stag Beetles that can be 70–80mm long.

The main characteristic of the order is the presence of tough, hardened forewings, known as elytra, which cover the membranous hindwings and usually meet closely down the midline of the abdomen. Beetles have biting jaws, which are adapted in different species to deal with a wide range of food types. All of them undergo complete metamorphosis, with their life-cycle consisting of egg, larva, pupa and adult.

Heath Tiger Beetle on heathland.

Green Tiger Beetle
Cicindela campestris

Tiger beetles are a large group of predatory, fast-moving beetles in the Carabidae family. This attractive and distinctive species is up to 16mm long. From above it appears wholly metallic bluish-green, with a few scattered cream-yellow spots on the wing-cases. The legs and underside are bristly-hairy. The beetle lives mainly on the ground in warm, sandy habitats, flying short distances with a rapid buzzing flight when necessary. The larvae live in holes in the ground, where they catch ants and other hapless insects.

FLIGHT PERIOD March–October.

HABITAT AND DISTRIBUTION Widespread and locally common in warm, sandy places such as heathland and sand-dunes.

SIMILAR SPECIES Heath Tiger Beetle *C. sylvatica* (also known as Wood Tiger Beetle) is very similar, but blackish-grey with four creamy spots and two zigzag lines on the wing-cases. It is uncommon, and occurs only in the south.

Green Tiger Beetle.

Violet Ground Beetle

Carabus violaceus

Members of the Carabidae family with around 350 species in Britain, ground beetles are voracious predators. This large, fast-moving species is up to 30mm long. The head, thorax and wing-cases are all metallic blue-purple, sometimes becoming coppery towards the flanged edges. The beetles are voracious nocturnal predators, hiding under stones and logs during the day, and emerging at night to seek out slugs and other invertebrates. If handled, they emit an unpleasant and persistent smell.

FLIGHT PERIOD All year.

HABITAT AND DISTRIBUTION Widespread and quite common throughout the area in various habitats, including woodland, parks and gardens.

SIMILAR SPECIES
C. problematicus is almost identical, except that it has much more ridged and wrinkled wing-cases. It is widespread but probably less common. *C. nemoralis* is slightly smaller and more bronze-green, with very ridged and pitted elytra. Its habitats and distribution are similar.

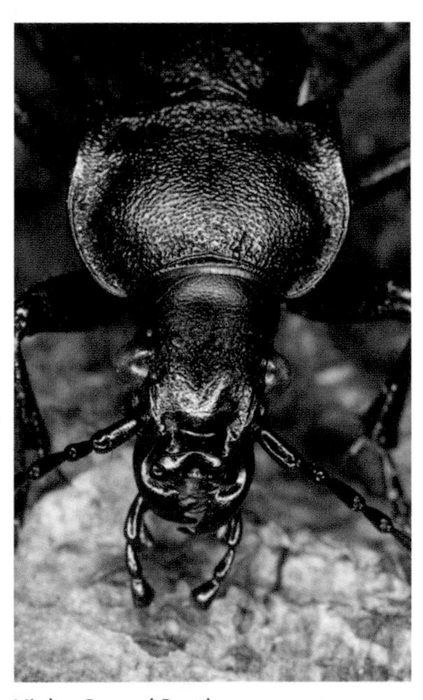

Violet Ground Beetle.

Burying Beetle

Nicrophorus interruptus

The burying beetles are a small group of insects (family Silphidae), also known as sexton beetles, which are highly adapted to finding and burying carcases of birds and mammals. This species has strikingly marked wing-cases with large orange patches on black, a black head and thorax, and large, orange-tipped, clubbed antennae. These allow the beetles to detect carrion, which they then set about burying, while also laying eggs in it, eating it and preying on other insects that visit it. Burying beetles are often covered with brownish mites, giving rise to another name, 'the lousy watchman'.

FLIGHT PERIOD March–October.

HABITAT AND DISTRIBUTION Widespread and locally common almost anywhere they can find carrion; they move around considerably at night and can turn up anywhere.

SIMILAR SPECIES There are several close relatives, differing in shape, elytra pattern and colour of the antennae.

A Burying Beetle, *Nicrophorus interruptus*.

Stag Beetle

Lucanus cervus

Stag Beetles belong to the large family Lucanidae, of which there are three species in Britain. The male is quite unmistakeable: at 75mm or so in length, it is Britain's longest beetle, with a bulky body and a pair of long, forward-projecting 'antlers'. The wing-cases are black or dark chestnut-brown, the head is blackish and the antlers are shiny reddish-black. The antennae are long with comb-like projections at the ends. Females are similar in colour, but smaller and lack the antlers.

The beetle's eggs are laid in substantial dead wood such as old tree trunks and large wooden posts. The larvae then take about four years to develop before the adults force their way out. Despite their appearance Stag Beetles are quite harmless to humans; the antlers are used in occasional fights between males. In flight they are usually seen around dusk, and are unmistakeable due to their large size, the odd angle the body is held at and their rather bumbling flight. They are particularly poor at landing.

Male Stag Beetle.

FLIGHT PERIOD June–July.

HABITAT AND DISTRIBUTION Locally common in areas where there is ample dead wood, especially in the New Forest and parts of Surrey, but generally rare and found only in the south.

SIMILAR SPECIES Lesser Stag Beetle *Dorcus parallelipipedus* is smaller, up to 30mm in length, all black and has much smaller antlers. It is quite common and widespread in the south in similar habitats, but is rarely seen as it is nocturnal.

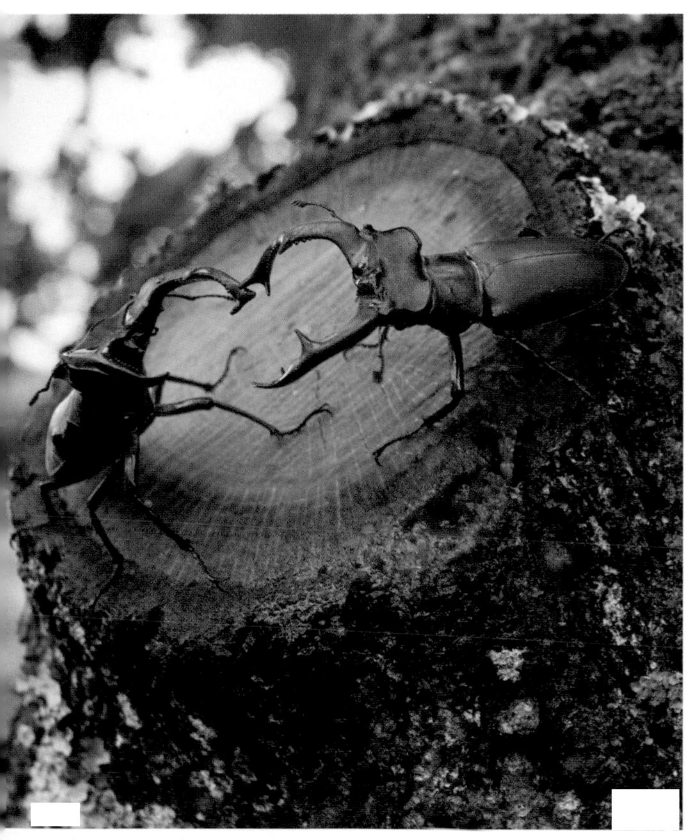

Male Stag Beetles in combat.

Dung beetles

Insects of the family Scarabaeidae, most dung beetles feed on dung. They fulfill a useful role in agriculture, burying and consuming dung and thus improving nutrient recycling and soil structure.

Minotaur Beetle
Typhaeus typhoeus

Male Minotaur Beetles are striking-looking creatures about 20mm long and shiny black. They have three projecting horns at the front, consisting of two long side horns and a shorter central one, more reminiscent of a rhinoceros than a minotaur. Females are slightly smaller, with two bumps in place of horns. Minotaur Beetles spend their working lives burying dung deep in the ground, then laying eggs in it. The species is most frequent on rabbit dung, which it rolls around to get it in the right position, although it also uses other dung.

FLIGHT PERIOD Active all year, although most often seen in spring and autumn.

HABITAT AND DISTRIBUTION Locally common in sandy habitats throughout England, Wales and Ireland, but absent further north.

SIMILAR SPECIES
The dor beetles *Anoplotrupes* spp. are similar in size and shape, but lack any horns (see opposite).

A Minotaur Beetle on rabbit dung.

Woodland Dor Beetle

Anoplotrupes stercorosus

The dor beetles particularly favour cow dung. The males and females together dig vertical shafts below the dung, then the female digs horizontal shafts from these in which to lay the eggs. The developing larvae feed on the dung. The insects are shiny black, roughly oval in outline and up to 25mm long, with a smooth pronotum – apart from fine puncture marks – and finely ridged elytra. They are bluer underneath. In common with other dung beetles, this species is often infested with mites.

FLIGHT PERIOD April–October.

HABITAT AND DISTRIBUTION Locally common throughout the area in cattle pastures.

SIMILAR SPECIES The dor beetle *Geotrupes stercorarius* is very similar, differing in minor details such as having 1.5 ridges instead of one on the outer side of the hind tibia.

Woodland Dor Beetle.

Chafers

Chafers belong to the Scarabaeidae family. Adults feed on the leaves and flowers of many plants, while the larvae live in soil and feed on plant roots.

Cockchafer or Maybug
Melolontha melolontha

Cockchafer is a large, noisy and distinctive beetle up to 30mm long. It has a black head and thorax, brown, grooved and sculptured wing-cases, and a pointed abdomen, and is very hairy below. The antennae are very attractive, with fan-shaped combs at the ends. Its alternative name of Maybug derives from its normal date of emergence. Cockchafers fly in the evenings and at night. They are readily drawn to lights and lit windows, which they may noisily crash into. Adults feed on leaves, while the larvae feed on the roots of plants in pasture. They were once so abundant that they were considered a pest, but have recently become steadily less common.

FLIGHT PERIOD May–June.

HABITAT AND DISTRIBUTION Locally common in grassy places and open woodland in England and Wales, but absent further north.

SIMILAR SPECIES Summer Chafer *Amphimallon solstitialis* is slightly smaller, has a brown thorax and is hairier. It flies in June–July, and is found in southern areas.

Cockchafer.

Garden Chafer

Phyllopertha horticola

This is a small chafer, up to 12mm long, which would be inconspicuous were it not for the fact that it can occur in some abundance at times. The head and thorax are greenish-black with scattered hairs, and the wing-cases are shiny chestnut-brown. It is very hairy underneath, like most chafers. It is a day-flying beetle by preference, and can be very abundant in favoured locations, sunbathing and feeding on low vegetation. The species can be a pest of fruit bushes at times.

FLIGHT PERIOD June–July.

HABITAT AND DISTRIBUTION Locally common throughout the area, especially on lighter soils in a variety of grassland and scrub habitats.

SIMILAR SPECIES None.

Garden Chafers on bracken.

Rose Chafer

Cetonia aurata

Rose Chafer is a beautiful jewel-like beetle that is bright shiny metallic green in colour and 15–20mm long. All the upper surface is bright metallic green (or sometimes bronzed or black), and there are short white transverse markings on the elytra, as well as a rather wavy rear edge to the elytra. The adults are day flying and spend much of their time visiting flowers of plants such as umbellifers and Elder to eat pollen and nectar. The larvae develop in dead wood, and may have an association with wood ants.

FLIGHT PERIOD May–August.

HABITAT AND DISTRIBUTION Locally common in flowery places near woodland, but absent from Scotland.

SIMILAR SPECIES *C. cuprea* is very similar, but has a less wavy edge to the elytra; it is found mainly in Scotland. Noble Chafer *Gnorimus nobilis* has distinctly wrinkled elytra, and more of a gap between the thorax and abdomen; it is rare.

Rose Chafer on Hawthorn blossom.

Bee Chafer or Bee Beetle

Trichius fasciatus

In Britain these beetles are unlikely to be mistaken for anything else. They are quite large, up to 25mm long, with a broad, rectangular body. The head and thorax are black underneath, but are densely covered by golden-brown hairs all over. The elytra are variably cream to orange, with black stripes and patches. The beetles are also very hairy underneath; the general impression is of considerable hairiness, almost as in a bumblebee. In addition, they have a heavy, buzzing flight, which heightens the resemblance. The larvae develop in rotten wood, and the adults are most commonly seen on flowers.

FLIGHT PERIOD May–August.

HABITAT AND DISTRIBUTION Found in flowery places, especially close to or in woodland; absent from south-east Britain and Ireland. Most common in Scotland.

SIMILAR SPECIES None.

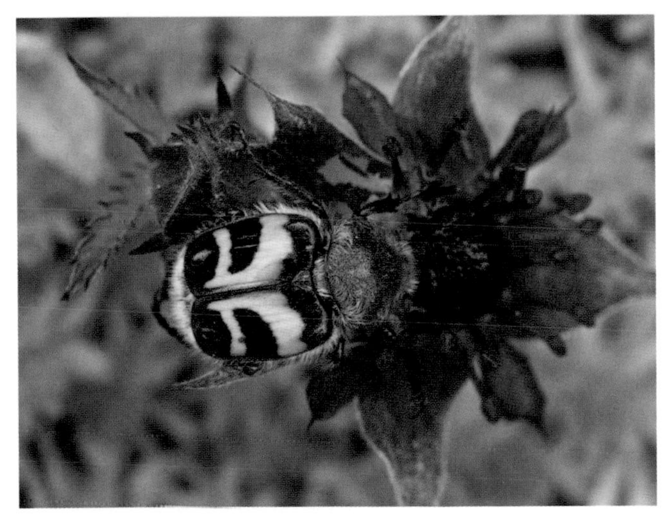

Bee Chafer on Marsh Cinquefoil.

Click Beetle

Athous haemorrhoidalis

The click beetles are quite a large family of insects, the Elateridae, with 73 species in the UK, although many are rare or inconspicuous. They are called click beetles because they have the ability to suddenly leap into the air with a loud click, either to escape danger or to turn themselves over. This species is probably the most common click beetle and the most often seen. It is 10–15mm long, with a hairy, grey-black thorax covered with shallow indentations, and brown, hairy elytra that are strongly grooved. The larvae, known as wireworms, are long and thin and can cause damage to crops.

FLIGHT PERIOD May–August.

HABITAT AND DISTRIBUTION Widespread throughout the area in a wide variety of grassy places, including cultivated land.

SIMILAR SPECIES *Agriotes lineatus* is similar but much smaller.

Click Beetle feeding on pollen.

Soldier beetles

The Cantharidae are a small family of attractive predatory beetles known generally as soldier beetles because of their smart, clearly defined colour scheme. They are narrowly cylindrical, slightly hairy and up to about 13mm long.

Soldier Beetle

Cantharis rustica

Up to 14mm long, this species has a blackish head, an orange-red pronotum, or thorax, with a clearly defined black spot in the middle, which does not reach the margins, and black elytra covered with downy hair. The beetles are most often seen at flowers, predating on visiting insects rather than taking nectar or pollen.

FLIGHT PERIOD May–August.

HABITAT AND DISTRIBUTION Widespread in southern Britain in a wide variety of grassy and flowery places.

SIMILAR SPECIES *C. fusca* is very similar, but the black spot on the pronotum, if present, reaches the front margin, and the legs are all black. It is found in similar habitats and is less common.

Soldier Beetle, *Cantharis rustica*.

Common Red Soldier Beetle

Rhagonycha fulva

Although this pretty little beetle is quite small, no more than 10mm long, it becomes conspicuous due its sheer abundance at times. The body is narrowly cylindrical with an orange-red head and thorax, and similarly coloured elytra, except that the distal quarter is black. Because of their general bright colouring the beetles are also known as bloodsuckers, although they are quite harmless to humans. They may be very abundant in flowery places, especially where there are umbellifers such as Wild Parsnip or Hogweed, and here they take both nectar and pollen, and predate visiting small insects. They are frequently seen as mating pairs.

FLIGHT PERIOD June–August.

HABITAT AND DISTRIBUTION Widespread in southern Britain in a wide variety of grassy and flowery places; less common in Scotland and Ireland.

SIMILAR SPECIES *Cantharis livida* is larger and does not have black-tipped elytra.

Mating Common Red Soldier Beetles.

Cardinal Beetle

Pyrochroa coccinea

This large and conspicuous beetle, one of three species of cardinal beetle (family Pyrochroidae) in the area, is rather flattened in appearance, up to about 20mm long, and has bright red elytra and thorax, and a black head. The name cardinal beetle derives from its bright red colour. The larvae are associated with dead and dying timber, but as predators on other invertebrates rather than as wood eaters. The adults visit flowers for pollen and nectar, or rest on foliage.

FLIGHT PERIOD May–July.

HABITAT AND DISTRIBUTION Widespread but local in southern Britain, in flowery habitats around woodland and old hedges.

SIMILAR SPECIES *P. serraticornis* has a red head and is otherwise similar in regard to both habits and habitats. Net-winged beetles such as *Platycis cosnardi* are similar but smaller, with less shiny, ridged wing-cases. They are much less common.

Cardinal Beetle, *Pyrochroa coccinea*.

Glow-worm

Lampyris noctiluca

Most people have heard of glow-worms, but few have seen them. Those found in Britain are in the family Lampyridae (fireflies). They are not a worm, but a type of beetle, although hardly a typical one. The males are like a normal beetle, 10–12mm long, with greyish-brown wing-cases and a dull brown head; they are able to fly well. The females, however, are wingless and look more like woodlice or larvae than beetles. They sit in grassy places and emit a greenish glow from the tip of the abdomen to attract the males which fly about in search of them.

When the larvae are mature they are very similar to females, differing in having pale orange spots at the rear of each segment, and in being more likely to be moving, in search of prey. They also glow, but less than adult females. The larvae are predators of snails, especially *Cepaea* species, which they paralyse and suck dry. Dead, undamaged snail shells on the ground in grassy places are a good indication of the presence of glow-worms.

Glow-worms are inconspicuous and rarely seen by day. They are most likely to be spotted as night falls, in unspoiled grassy places, especially but not necessarily on calcareous soils; let your eyes adjust to the darkness and look for the lights. They can be common in good places, especially away from the lights of houses and streets.

FLIGHT PERIOD May–August.

HABITAT AND DISTRIBUTION Widespread but local in southern Britain in a wide variety of grassy and flowery places; less common in Scotland and probably absent from Ireland.

SIMILAR SPECIES The very rare Lesser Glow-worm *Phosphaenus hemipterus* is much smaller and does not glow.

Right: Snail killed by Glow-worm.

Glow-worm larva.

Ladybirds

Ladybirds (family *Coccinellidae*) are familiar and popular insects known to almost everyone. The best-known one is 7-spot Ladybird, but there actually 46 species in the UK, all with their own characteristics.

Ladybirds are typically roughly hemispherical or half-ovoid, markedly domed and brightly coloured, with the head being almost concealed by the hardened pronotum. Many have variable numbers of spots. 7-spot and some of the other species are carnivorous as both adults and larvae, feeding voraciously on aphids; this makes them popular with gardeners and farmers. The bright colours of ladybirds are a warning colouration to potential predators, indicating that they taste unpleasant, so they tend not to be predated. All the British species are active in warm weather and hibernate as adults.

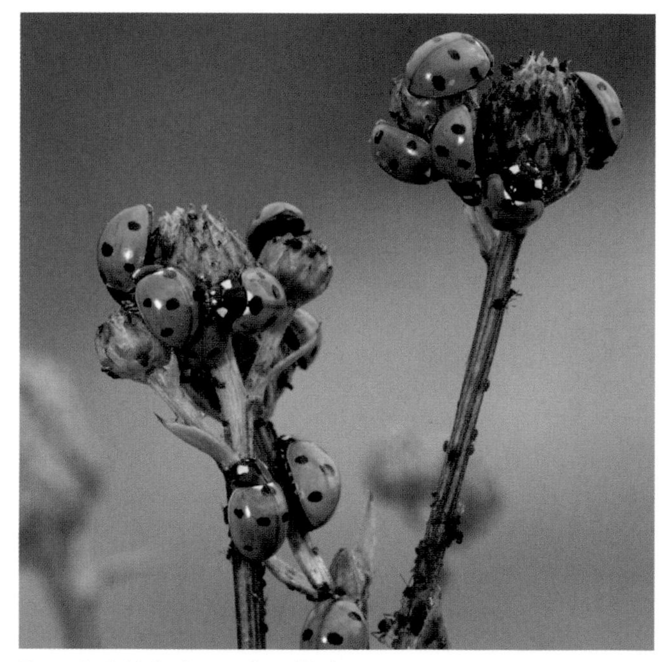

7-spot Ladybirds clustered on thistles.

7-spot Ladybird *Coccinella 7-punctata* is about 8mm long and has bright red elytra with seven black spots scattered over them, and a black pronotum with two white spots. It is common throughout the area in almost any habitat, and large numbers may arrive from mainland Europe in favourable conditions.

Harlequin Ladybird *Harmonia axyridis* was first seen in Britain in 2004 and has since spread at an astonishing rate, often ousting native species. It is hugely variable and sometimes hard to identify. It is roughly the same size and shape as 7-spot, but most commonly the elytra have 15–21 black spots on orange, or two or four red spots on black; it almost never has exactly seven spots. The pronotum is black with large variable white spots, and the legs are always brown (compared with black in 7-spot). Harlequin Ladybirds are now abundant almost everywhere and hibernate in masses.

Clustered Harlequin Ladybirds in winter.

Cream-spot Ladybird.

Cream-spot Ladybird *Calvia 14-guttata*, also known as 14-spot, is orange or brown with seven spots on each wing-case. It is common and widespread.

Eyed Ladybird *Anatis ocellata* is one of the largest species, up to 10mm long, and distinctive in having white eye-rings around each of the 18 or so black spots. It is common and widespread throughout the UK in conifer woods.

An Eyed Ladybird.

Thick-legged Flower Beetle
Oedemera nobilis

This is a common and distinctive beetle (family *Oedemeridae*) with various other common names, all referring to the greatly enlarged hind legs of the male. The beetles are up to 11mm long and bright shiny green, with wing-cases that become progressively further apart along the midline down the thorax. Males have greatly enlarged hind thighs that are almost bulbous, whereas females do not. The larvae live in the hollow stems of various plants, and the adults visit a wide variety of flowers.

FLIGHT PERIOD April–August.

HABITAT AND DISTRIBUTION Widespread in southern Britain only, in flowery habitats, especially around woodland.

SIMILAR SPECIES *O. lurida* is similar in shape but a little smaller and, despite the name, rather duller in colour. The legs of the male are only slightly swollen. Its habitats and distribution are similar.

Male Thick-legged Flower Beetle.

Oil Beetle

Meloe proscarabeus

The oil beetles are a small family (Meloidae) of largish beetles with an extraordinary life-cycle and distinctive appearance. There are about 10 species in the UK, but several are very rare and others have become extinct in recent times.

This species is the one most often seen and is typical of the group. The adults are shiny bluish-black, with very short wing-cases (they are flightless) and a swollen abdomen especially in the female. Females are up to 30mm long, males rather less. After mating in spring females dig burrows in banks near solitary bee nests, in which they lay large quantities of eggs. These hatch into distinctive active, legged larvae known as triungulins (occasionally spelt as tringulins). They in turn emerge to attach themselves to solitary bees, which take them back to their nests. There they change to grub-like larvae and begin feeding on the eggs, larvae and pollen of the host. The larvae of some species may emerge and remain as a mass, and there is some evidence that this mass mimics the shape and scent of the female target bee.

Female Oil Beetle.

Mating Oil Beetles.

 **ADULT PERIOD** April–July.

HABITAT AND DISTRIBUTION Locally common in unimproved warm, flowery places where the host bees occur in abundance, throughout southern Britain and Ireland.

SIMILAR SPECIES The other species are less common. *M. variegatus* is metallic blue-green and has a definite multi-coloured rainbow mark on each abdomen segment; *M. violaceus* is bluer, with finer dotting on the head and thorax.

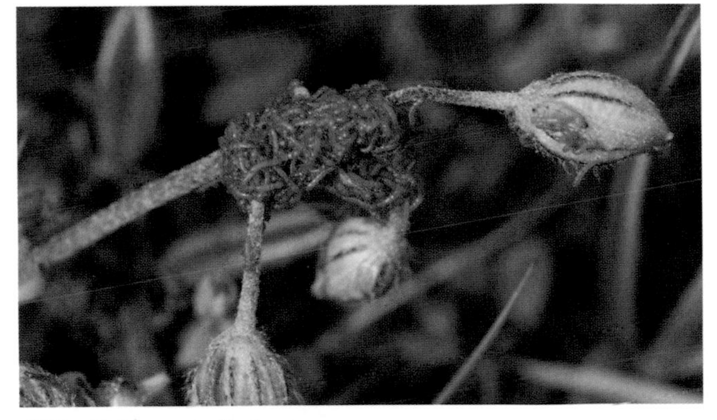

Oil Beetle larvae on Rock-rose.

Longhorn beetles

This is a very large family (Cerambycidae) of beetles worldwide, represented by about 40 species in Britain. They are very distinctive, always having long antennae (normally longer in the male), often being large in size and having long, narrow, flattened bodies. The larvae develop in dead and dying wood, so most species are associated with old woodland, especially where there are flowers for the adults to feed on. They are almost all very attractive species that are always a pleasure to find, particularly as few of them are common.

Spotted Longhorn *Ruptela maculata* is one of the most often-seen longhorns. It is 15–20mm long with a black head and thorax, and dirty yellow elytra marked with four black bands or blotches. It can be seen visiting flowers in May–August, in woodland habitats throughout the area.

Mating Spotted Longhorn beetles.

Wasp Beetle.

Wasp Beetle *Clytus arietis* is rather similar, but smaller and more slender, to 15mm long, with an essentially black body marked with yellow stripes on the wing-cases, giving a reasonable impression of a wasp. The larvae live in various trees and even fence posts. Flight occurs in May–August in flowery places near trees.

The wonderfully named **Golden-bloomed Grey Longhorn** *Agapanthia villosoviridescens* is a large and distinctive species, up to 25mm long, with two marked golden stripes on the thorax, a golden bloom all over the greyish elytra and very long, chequered and hairy antennae. It is an uncommon species in the UK, found mainly in eastern England, and is much more common in mainland Europe. It flies in May–July in damp, flowery places, especially those associated with thistles.

Golden-bloomed Grey Longhorn beetle.

Musk Beetle

Aromia moschata

This very striking large longhorn beetle is up to 35mm long, with a long, slender body that is wholly metallic bronze-green in colour, although it may also be blue-green, black or coppery, and long antennae. The shape, size and metallic sheen are distinctive. The larvae live in old but not necessarily rotten willows and poplars, while the adults spend much of their time eating pollen on umbellifers such as Hogweed and Angelica.

FLIGHT PERIOD June–September.

HABITAT AND DISTRIBUTION Widespread but local in England, Wales and Ireland, in flowery habitats, especially where there are willows or poplars.

SIMILAR SPECIES No longhorns are similar; Spanish Fly Beetle *Lytta vesicatoria* has a similar shape and colour, but is less than 20mm long with much shorter antennae. It is rare and found in southern locations only.

Musk Beetle on Angelica.

House Longhorn Beetle

Hylotrupes bajulus

This medium-sized and very variable beetle is 10–22mm long. It is quite hairy, especially when newly emerged. Its colour varies from black marked with cream to brown, but it always has two shiny, round black bumps on the thorax. Also known as Old House Borer, it is a noted pest of building timbers, telegraph poles and other dead softwoods, which the larvae eat voraciously.

FLIGHT PERIOD May–August.

HABITAT AND DISTRIBUTION Local, and found mainly in the south-east and Midlands of England, with scattered records elsewhere. Usually associated with built-up areas.

SIMILAR SPECIES None.

House Longhorn Beetle on timber.

Leaf-beetles

This is a large family (Chrysomelidae) of herbivorous beetles that are generally rounded in shape, shiny and brightly coloured. There are about 270 species in Britain, including a number of pest species.

Lily Beetle
Lilioceris lilii

Although small, up to 8mm long, Lily Beetle is highly conspicuous by virtue of its bright, shiny red thorax and elytra, and by its abundance. It is an abundant garden and horticultural pest of lilies and fritillaries, and may also live on other members of the lily family. The larvae cover themselves with their own slimy excrement for protection.

FLIGHT PERIOD April–September.

HABITAT AND DISTRIBUTION
Widespread wherever there are host plants; it was introduced into the UK in 1943 and has spread almost throughout the area.

SIMILAR SPECIES
May be confused with cardinal beetles (see p. 165), but they are larger and have comb-shaped antennae.

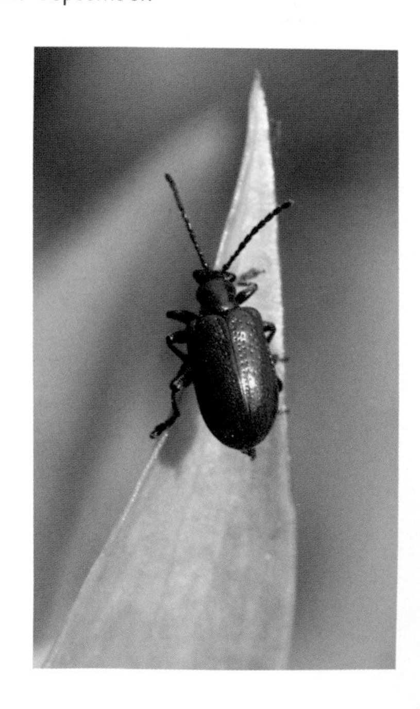

Beetle on lily.

Bloody-nosed Beetle

Timarcha tenebricosa

This is the largest of the leaf-beetles in the UK, reaching 20mm in length. It is a domed, dark bluish-black beetle that is unable to fly, with the elytra fused together. Its feet look unusually large. It derives its name from a defence mechanism – it emits a few drops of bright red fluid from its mouth when threatened; this is said to put off bird predators. The larvae, which are blackish, rounded and look rather inflated, feed on various bedstraws, especially Hedge Bedstraw.

FLIGHT PERIOD April–August.

HABITAT AND DISTRIBUTION Widespread but local throughout the area in warm, open habitats such as chalk downland and coastal grassland.

SIMILAR SPECIES Small Bloody-nosed Beetle *T. goettingensis* is smaller, to 13mm in length, with a pronotum (thorax) that does not angle sharply inwards to form a 'waist'. It has similar habits, and is less common.

Bloody-nosed Beetle.

Mint Beetle
Chrysolina herbacea

This is a striking, bright iridescent green, rounded beetle up to 10mm long, with finely punctured elytra. Both the fat black larvae and adults feed on mint, leaving holes in the leaves, although they are rarely considered a problem in gardens. Mint Beetle was formerly known as *C. menthastri*, and is also known as Green Mint Beetle to distinguish it from the recently introduced Blue Mint Beetle *C. coerulans*.

FLIGHT PERIOD May–September.

HABITAT AND DISTRIBUTION Widespread in damp riverside habitats and marshy fields where mint is abundant. Rarer in the north.

SIMILAR SPECIES There are similar beetles, such as *Cryptocephalus hypochaeridis*, but not on mint.

eetle on mint.

Colorado Beetle
Leptinotarsa decemlineata

This striking and unmistakeable beetle is rounded, domed and up to 10mm long. The head and thorax are orange-brown marked with black blotches, and the elytra are boldly striped with black on pale yellow. The larvae are fat and red spotted with black. Colorado Beetles are a notorious and damaging pest of potatoes, and less commonly tomatoes. They are rarely seen in Britain, but as a notifiable pest they have to be reported if seen.

FLIGHT PERIOD April–September, but may emerge from hibernation at other times.

HABITAT AND DISTRIBUTION Common in continental Europe wherever the host plants occur in quantity. Rare occasional visitor in Britain.

SIMILAR SPECIES None.

Colorado Beetle.

Pine Weevil

Hylobius abietis

The weevils (family Curculionidae) are a very distinctive group of smallish beetles with their own characteristics, but are often hard to identify to species from among the 475 or so UK species. The front of the head of a weevil is prolonged into a narrow beak, or rostrum, with jaws at the tip and elbowed antennae inserted partway along it. Most weevils are plant eaters and many are flightless, with their wing-cases fused together. Pine Weevil is one of the larger (up to 13mm long) and most common weevils. It is generally black in colour and covered in irregular patches of golden-yellow or cream hairs, with a thick, black, curved rostrum. In some places it can be a serious pest of coniferous forests, where it chews the new shoots.

Pine Weevil.

FLIGHT PERIOD Adult all year, but most often seen April–September.

HABITAT AND DISTRIBUTION Widespread and common throughout the area in pine and other coniferous forests.

SIMILAR SPECIES Vine Weevil *Otiorhynchus sulcatus* is slightly smaller and paler with longer antennae. It is a garden pest. Yellow Weevil *Lixus paraplecticus* is a striking species, up to 16mm long, slender and dark grey in colour, but covered with golden yellow scales. It is common on the Continent in damp places with umbellifers, but is probably now extinct in Britain.

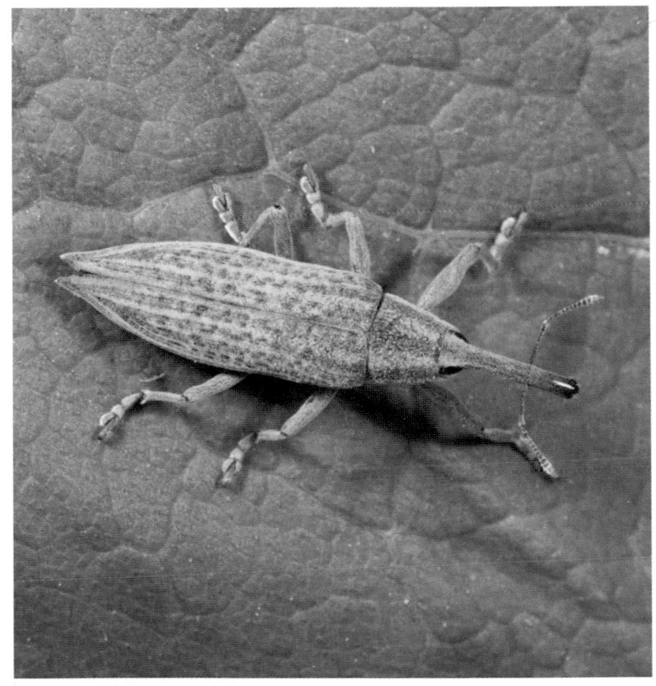

Yellow Weevil *Lixus paraplecticus*.

Whirligig Beetle
Gyrinus substriatus

Whirligig beetles (family Gyrinidae), of which there are several species in the UK, are appealing aquatic insects that spend their time racing around the surface of still and gently moving water bodies in apparently aimless circles, using their legs as paddles. They have eyes modified into two parts to see both above and below the water line. They are predators, especially on mosquito larvae.

FLIGHT PERIOD Adult all year, but most often seen April–September.

HABITAT AND DISTRIBUTION Widespread and common throughout the area in clean, still and slow-moving waters such as ditches.

SIMILAR SPECIES There are several other species, but detailed examination is required to separate them. *G. natator* was formerly considered to be the most common species, but recent work has shown it to be quite rare.

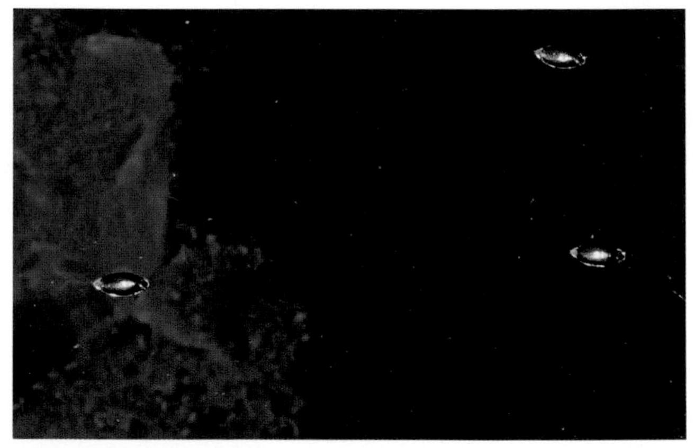

Whirligig beetles on pond.

SPRINGTAILS, COLLEMBOLA

Strictly speaking the springtails are no longer considered to be insects, but their exact evolutionary and taxonomic status is still subject to discussion.

Springtail
Anurida maritima

Most springtails live hidden in leaf litter or under stones, but this species is much more conspicuous, albeit small. Individuals are 2–3mm long, cylindrical and grey in colour, but they collect together in large groups of up to 100 or so, floating on the surfaces of rockpools. Unlike most springtails, they have no 'spring' because it would be of little value in an aquatic environment.

FLIGHT PERIOD Adult all year.

HABITAT AND DISTRIBUTION Widespread around the UK and Irish coasts in rockpools and rocks of the upper tidal zone.

SIMILAR SPECIES Water Springtail *Podura aquatica* is similar, but occurs on fresh water.

Springtails, *Ariunda maritima*, in rock pool.

GLOSSARY

abdomen: third, rear part of an insect.

antennae (singular antenna): paired, often long sensory appendages on the head of an insect, used for feeling and smelling.

cerci (singular cercus): paired appendages at the rear of the abdomen on some insects, such as the 'pincers' of an earwig, or the 'tails' of a stonefly.

elytra (singular elytron): hard wing-cases of beetles, not actually used in flight.

endemic: restricted in distribution to a specific, often small area.

halteres: tiny, pin-shaped balancing organs that replace the hind-wings in true flies.

larva (plural larvae): stage of the life-cycle of insects that undergo complete metamorphosis, which concentrates particularly on feeding; also called caterpillar.

metamorphosis: all insects undergo big changes in their lives. Those with complete metamorphosis have the stages egg – larva – pupa – adult. Those with so-called partial metamorphosis (such as grasshoppers) have an egg, followed by a succession of gradually maturing nymphs, and finally a fully winged, sexually mature adult.

nymph, nymphal: see metamorphosis.

ovipositor: egg-laying organ of a female insect, used to position the egg. It can be hidden or very conspicuous, as in bush-crickets.

parthenogenesis: reproduction without fertilisation.

pronotum: dorsal (upper-side) surface of the first segment of the thorax, particularly significant in some groups, such as beetles.

pterostigma: coloured cell near the wing-tip of dragonflies and some other insects.

stridulation: creation of a sound by rubbing two body parts together, as in bush-crickets.

thorax: second part of an insect, immediately behind the head.

BIBLIOGRAPHY AND RESOURCES

British Wildlife Magazine, www.britishwildlife.com/themagazine. An excellent magazine on all aspects of British nature.

Brock, Paul D. *A Photographic Guide to Insects of the New Forest*. Pisces Publications, Newbury, 2011.

Brooks, Steve. *Field Guide to the Dragonflies and Damselflies of Great Britain and Ireland*. British Wildlife Publishing, Dorset, 2004.

Chinery, M. *Insects of Britain and Western Europe*, 3rd edn. Bloomsbury Publishing, 2012.

Chinery, M. *Collins Complete Guide to British Insects*. Collins, London, 2005.

Gibbons, Bob. *Field Guide to the Insects of Britain and Northern Europe*. Crowood Press, Wiltshire, 1995.

Sterling, Phil et al. *Field Guide to the Micro-moths of Great Britain and Ireland*. British Wildlife Publishing, Dorset, 2012.

Stubbs, Alan & Drake, Martin. *British Soldierflies and their Allies*. BENHS, Reading, 2001.

Stubbs, Alan & Falk, Stephen. *British Hoverflies, an Illustrated Identification Guide*, 2nd edn. BENHS, Reading, 2002.

Thomas, Jeremy. *The Butterflies of Britain and Ireland*, 2nd edn. British Wildlife Publishing, Dorset, 2010.

Waring, Paul et al. *Field Guide to the Moths of Great Britain and Ireland*. British Wildlife Publishing, Dorset, 2003.

ORGANISATIONS AND SOCIETIES

Amateur Entomologists' Society
www.amentsoc.org
Books, a magazine, events and so on.

Buglife
www.buglife.org.uk
Join to help conserve insects, or visit the website for species identification, surveys and campaigns.

British Dragonfly Society
www.british-dragonflies.org.uk
Publications, events, recording schemes and so on.

Butterfly Conservation
www.butterfly-conservation.org
Helping to conserve and learn about butterflies and moths. Lots of local events.

Bees, Wasps and Ants Recording Society
www.bwars.com
Everything to do with hymenopterans.

Field Studies Council
www.field-studies-council.org
Courses on all aspects of nature and the environment, including insects.

The Kingcombe Centre, Dorset
www.kingcombe.org
Various courses and accommodation in the middle of a lovely insect-rich area.

The Wildlife Trusts
www.wildlifetrusts.org
The umbrella organisation for all the county Wildlife Trusts. They play a crucial role in conserving habitats for insects and all other wildlife throughout the UK.

INDEX

INDEX

INDEX